T0339608

Assessing the Energy Efficiency of Pumps and Pump Units

Assessing the Energy Efficiency of Pumps and Pump Units

Background and Methodology

Bernd Stoffel
Professor (retired)
Darmstadt University of Technology
Germany

ELSEVIER

AMSTERDAM • BOSTON • HEIDELBERG • LONDON
NEW YORK • OXFORD • PARIS • SAN DIEGO
SAN FRANCISCO • SINGAPORE • SYDNEY • TOKYO

Elsevier
Radarweg 29, PO Box 211, 1000 AE Amsterdam, Netherlands
The Boulevard, Langford Lane, Kidlington, Oxford OX5 1GB, UK
225 Wyman Street, Waltham, MA 02451, USA

Copyright © 2015 Elsevier Ltd. All rights reserved.

No part of this publication may be reproduced or transmitted in any form or by any means, electronic or mechanical, including photocopying, recording, or any information storage and retrieval system, without permission in writing from the publisher. Details on how to seek permission, further information about the Publisher's permissions policies and our arrangements with organizations such as the Copyright Clearance Center and the Copyright Licensing Agency, can be found at our website: www.elsevier.com/permissions

This book and the individual contributions contained in it are protected under copyright by the Publisher (other than as may be noted herein).

Notices
Knowledge and best practice in this field are constantly changing. As new research and experience broaden our understanding, changes in research methods or professional practices, may become necessary.

Practitioners and researchers must always rely on their own experience and knowledge in evaluating and using any information or methods described herein. In using such information or methods they should be mindful of their own safety and the safety of others, including parties for whom they have a professional responsibility.

To the fullest extent of the law, neither the Publisher nor the authors, contributors, or editors, assume any liability for any injury and/or damage to persons or property as a matter of products liability, negligence or otherwise, or from any use or operation of any methods, products, instructions, or ideas contained in the material herein.

ISBN: 978-0-08-100597-2

British Library Cataloguing-in-Publication Data
A catalogue record for this book is available from the British Library

Library of Congress Cataloging-in-Publication Data
A catalog record for this book is available from the Library of Congress

For Information on all Elsevier publications
visit our website at http://store.elsevier.com/

This book has been manufactured using Print On Demand technology.

Working together
to grow libraries in
developing countries

ELSEVIER | Book Aid International

www.elsevier.com • www.bookaid.org

Contents

About the Author

 Professor Dr. Bernd Stoffel was head of the Chair for Turbomachinery and Fluid Power at the Darmstadt University of Technology (Germany) from 1985 until his retirement in 2006. Previous stations of his professional career included being a research assistant at the Karlsruhe University of Technology (Germany) and an executive employee at the German pump manufacturer KSB. His research work and experience cover various aspects of turbomachines, such as design, operation, and efficiency of pumps, and is documented by many scientific or application-oriented publications in journals and on conferences. For the past several decades, he has been involved in various national and international activities and working groups of the German Association of Pump Manufacturers (VDMA) and of Europump.

Preface by the President of Europump

Dear reader,

In front of you lies the first comprehensive handbook about the *savoir-faire* of pumps and pump units that is authored by Europump. This is an easy-to-read book that will be appreciated by students and teachers in academia, engineers, industrial plant operation managers, policymakers, and pump manufacturers.

Ecopump is the name of the successful flagship project that Europump launched in 2004 to bring the pump industry's commitment to energy savings to the fore. The need for scientific demonstration of potential energy savings is still the departure point for any legislative measure affecting the pump industry.

The future shines bright on Ecopump, too. Written in astonishingly clear language, *Assessing the Energy Efficiency of Pumps and Pump Units—Background and Methodology* gives a full update with new insights and new statistics. Europump owes it to long-standing cooperation with the Darmstadt University of Technology and in particular the writer, Professor Dr. Bernd Stoffel.

I wish you happy reading,

Carlo Banfi
President Europump 2013−2015

Acknowledgments

The author wishes to express his thanks to Europump for getting him the occasion to be involved in the common work on energy efficiency assessment; for contracting with him for the elaboration of methods, results, proposals for standards; and last but not least, for the writing of this book. The contributions of the author to the common work as well as the various results achieved by the common work and presented in this book are based on the fruitful cooperation of outstanding experts from several manufacturing companies in various European countries. The progress of the activities of the relevant Europump Joint Working Group and its subgroups has ever been (and still continues to be) steered and enforced by the chairman Dr. Horst-Georg Schmalfuss with great emphasis and personal involvement. The author is particularly indebted to Markus Teepe for critically looking through raw versions of several chapters of the manuscript of this book and for giving valuable comments and suggestions for improvements.

Members EUROPUMP JWG EuP

Svend Amdisen	Grundfos Management A/S	Denmark
Niels Bidstrup	Grundfos Management A/S	Denmark
Heinz Bohn	Grundfos Management A/S	Denmark
Frank Lorentzen	Grundfos Management A/S	Denmark
Ole Lund Markussen	Grundfos Management A/S	Denmark
Laurent Aillerie	Wilo Salmson France	France
Stephane Brousseau	Wilo Salmson France	France
Julien Chalet	PROFLUID	France
Robert Dodane	Wilo Salmson France	France
Alexandre Etienne	Wilo Salmson France	France
Jean-François Kerleroux	Wilo Salmson France	France
Gerhard Berge	KSB Aktiengesellschaft	Germany
Manfred Britsch	ALLWEILER GmbH	Germany
Michael Burghardt	Danfoss GmbH	Germany
Frank Ennenbach	Sulzer Management AG	Germany
Ingo Fabricius	WILO SE	Germany
Edgar Große Westhoff	WILO SE	Germany
Hansjürgen Kech		Germany
Thorsten Kettner	WILO SE	Germany
Friedrich Klütsch	VDMA	Germany
Peter Knapp		Germany
Michael Könen	KSB Aktiengesellschaft	Germany
Sebastian Lang	Technische Universität Darmstadt	Germany
Gerhard Ludwig	Technische Universität Darmstadt	Germany
Peter Pelz	Technische Universität Darmstadt	Germany
Horst-Georg Schmalfuß		Germany
Paul Taubert	Technische Universität Darmstadt	Germany
Markus Teepe	WILO SE	Germany
Lutz Urban	KSB Aktiengesellschaft	Germany
Peter Zwanziger	Siemens AG	Germany
John Bower	Flowserve Pump Ltd	Great Britain
Ken Hall	CALPEDA Limited	Great Britain
John Hollins	SPP Pumps Limited	Great Britain
Kathryn Poke	Armstrong Integrated Limited	Great Britain
Wayne Rose	Armstrong Integrated Limited	Great Britain
Steve Schofield	British Pump Manufacturers' Association	Great Britain
Carlo Anapoli	DAB Pumps Spa	Italy
Marco Bernacca	PENTAIR Water Italy	Italy
Matteo Cipelli	LOWARA srl	Italy
Michele Dai Pre	Calpeda s. p. a.	Italy
Andrè Prandoni Kistner	Calpeda s. p. a.	Italy
Fabio Reffo	LOWARA srl	Italy
Franco Strozzi	Caprari S.p.A.	Italy
Enrico Trentin	DAB Pumps Spa	Italy
Markus Holmberg	Xylem Water Solutions	Sweden
Fredrik Söderlund	Xylem Water Solutions	Sweden
Mehmet Kaya	Standart Pompa ve Mak. San. Tic. A.S.	Turkey

Abbreviations

AC	alternating current
BEP	best efficiency point
CDM	complete drive module (frequency inverter + control electronics)
CEN	European Standardization Committee
CENELEC	European Committee for Electrotechnical Standardization
DC	direct current
EC	European Commission
EEI	energy efficiency index
EMDS	electric motor-driven systems
EN	European Standard
EPA	extended product approach
EU	European Union
EUROPUMP	European association of pump manufacturers
EuP	energy-using products
HVAC	heating, ventilation, and air conditioning
IEC	International Electrotechnical Commission
IEx	efficiency level (=class) of a motor or a CDM
IESx	efficiency level of a PDS
ISO	International Organization for Standardization
JWG	Joint Working Group
MEI	minimum efficiency index
PA	product approach
PDS	power drive system (electric motor + CDM)
PM	permanent magnet
RCDM	reference complete drive module
RPDS	reference power drive system
RM	reference motor
SAM	semi-analytical model
TUD	Technische Universitaet Darmstadt (Darmstadt University of Technology)
VSD	variable-speed drive

List of Abbreviated Subscripts

aux	of auxiliaries
BEP	value at best efficiency point
corr	corrected
el	electric
hyd	hydraulic
i	consecutive number
imp	impeller
man	manufacturing
max	maximum
meas	measured
mech	mechanical
min	minimum
L	loss
M	of the motor
N	nominal
OL	overload
pL	related losses at the supporting load points
P	mechanical power
PL	part-load
r	random
r	rated
ref	reference value
requ	required
s	systematic or instrument
st	stages of multistage pumps
sync	synchronous
tot	total
T	in a time period T
x	of a quantity x

List of Symbols

c	flow velocity
c_T	torque friction coefficient
D	diameter
e	measurement uncertainty
E	energy
f	frequency
g	gravitational constant
H	head
i	consecutive number
i_{st}	number of stages
I	electric current
k, k^*	coefficient
k	surface roughness
L	length
n	rotational speed
n_s	specific speed
N	total number
p_L	related loss
P	power
P_L	compensated power of circulators
Q	flow rate
Re	Reynolds number
s	slip
S_{equ}	apparent electric power
SF	shape factor
t	time
t	tolerance
T	torque
T	time period
u	circumferential velocity
U	electric voltage
x	general variable
z	total number
η	efficiency
λ	friction coefficient
ν	kinematic viscosity
ω	angular velocity
ρ	density
σ	standard deviation

Introduction

In the first three chapters, this book tries to give an overview on the motivation for assessing the energy efficiency of pumps and pump units (**Chapter 1**), on the present state of corresponding standardization and legislation (**Chapter 2**), and on activities of Europump in respect to this issue (**Chapter 3**). The physical and technical background of pump efficiency and its dependencies and limitations are basically compiled in **Chapter 4**. The basic explanations are extended to the electric power input of motor-driven pumps in **Chapter 5**. The effect of manufacturing tolerances on the resulting tolerance of efficiency or efficiency indicators is addressed in **Chapter 6**. The Minimum Efficiency Index (MEI) is already in use for assessing pumps as separate products and for proving their compliance with a corresponding EU Regulation. The origins of the definition of MEI and the test method for its determination are shortly described in **Chapter 7**. The Energy Efficiency Index (EEI) is an appropriate indicator for assessing the energy efficiency of pumps together with their electric drives. EEI is already established for proving the compliance of circulators with a corresponding EU Regulation. Recent activities of Europump aimed at developing specific definitions and methods that are needed to apply the concept of EEI to pump units consisting of separate pumps and electric drives. The essentials of the concept of EEI are provided in **Chapter 8**.

Most of the various chapters can be read independently of each other. Chapter 1 serves to give valuable information on the global and European situation in respect to energy generation and consumption and particularly on possibilities and potentials to save energy consumed by pumping systems. Chapter 3 may be worth reading to member companies of Europump but also to national, European, and non-European institutions, associations, and authorities, as it gives compact information on recent and current activities of Europump relating to the topic of the book. Preferably, reading Chapters 2, 6, 7, and 8 is recommended to those who are mainly interested in getting the most relevant information on existing and expected assessing methods and on corresponding efficiency indicators and fundamental relations. Chapters 4 and 5 may be useful to read for those who are not so familiar with the physics behind efficiency and losses of pumps and electric drives or who want to refresh their knowledge in these fields.

Chapter 1

The Role of Pumps for Energy Consumption and Energy Saving

1.1 GENERATION AND CONSUMPTION OF ELECTRIC ENERGY

The utilization of electric energy for many various purposes is a characteristic aspect of modern technics and human life.

1.1.1 Generation of Electric Energy

Electric energy is generated by converting primary energy sources. A significant part of these primary sources consists of the fossil combustibles black coal, brown coal, mineral oil, and natural gas. These combustibles are converted into electric energy in conventional power plants. A second part of primary sources consists of radioactive materials, which are used in nuclear power plants to generate electric energy by nuclear fission. A third and increasing part of primary sources is renewable (water power, wind, solar radiation, biomass) and is converted in different technical facilities (e.g., water power plants, wind turbines, solar energy plants, biomass power plants) into electric energy.

In the European Union (EU), the total generation of electric energy was 3295 TWh in 2012 [1].

According to Ref. [2], the following mix of electric energy generation in the EU existed in 2007 and is expected for 2020, respectively (Table 1.1):

One aspect of both the fossil-fueled and the nuclear power plants is their contribution to the irreversible consumption of limited reserves of primary energy sources. This is a first motivation, to reduce electric energy generation by saving electric energy.

In the case of nuclear power plants, the risk of nuclear accidents and the issue of nuclear waste having possible effects on the environment and on humans are motivations to reduce their number worldwide.

Concerning the fossil-fueled power plants, a severe aspect is their emission of CO_2, which is classified as a "greenhouse gas" and is responsible for medium- and long-term climate change. Therefore, international commitments, especially

Assessing the Energy Efficiency of Pumps and Pump Units.
DOI: http://dx.doi.org/10.1016/B978-0-08-100597-2.00001-X
© 2015 Elsevier Ltd. All rights reserved.

TABLE 1.1 Mix of Electric Energy Generation in Europe

	2007	2020
By fossil-fueled power plants	56.0%	46.5%
By nuclear power plants	28.0%	21.0%
From renewable sources	16.0%	32.5%

in the EU and in its member countries, aim for the general reduction of CO_2 emissions by certain amounts and by certain dates. The major EU policy package, called the climate and energy package, which was adopted and became a binding legislation in 2009, includes as one of the targets for 2020 the reduction in EU greenhouse gas emissions of at least 20% below 1990 levels.

This leads—besides other goals, such as to reduce the CO_2 emission of traffic and heating of buildings—to the requirement of reducing CO_2 emissions that result from the generation of electric energy. Depending on the fossil fuels used as primary sources in the different types of fossil-fueled power plants, on their plant efficiency, and on the mix of the electric energy generated by them, the emission of CO_2 caused by generating electric energy in fossil-fueled power plants ranges from 630 to 980 g/kWh, according to Ref. [3]. This means that the mass of CO_2 emitted from fossil-fueled power plants is approximately proportional to the electric energy generated by them. In combination with the numbers given above, this results in an emission of approximately $(2-3) \cdot 10^9$ t of CO_2 by generation of electric energy in fossil-fueled power plants within the EU in 2012.

The electric energy generated in power plants and other generation facilities of different types and located at various sites is fed into common electric supply networks and transmitted to each end user. It is supplied to the end users as three-phase alternating current (AC) of constant voltage and frequency. In the EU, the latter is generally 50 Hz for residential, public, and industrial applications.

1.1.2 Consumption of Electric Energy

The total consumption of electric energy can be divided into several categories, such as driving (electric motors), lighting, heating, communication, information, and others.

Concerning the *worldwide* situation, it is estimated in Ref. [5] that electric motor driven systems (EMDSs) account for between 43% and 46% of the global electricity consumption. This amount is more than twice that of the second largest, which is lighting, contributing by 19% to the total consumption.

The share of electric energy consumption by motor-driven systems to the various sectors of application is given in Ref. [5] as:

Industry	64%
Commercial	20%
Residential	13%
Transport and Agriculture	3%

with a total consumption of about 7100 TWh/year. This value is expected in Ref. [5] to rise to more than 13,000 TWh/year by 2030 if no comprehensive and effective measures to improve the energy efficiency of motor-driven systems are taken soon.

The *global* consumption of electric energy by electric motors is dominated by four major motor applications. According to Refs [5,8], in 2006 the corresponding share was as follows: compressors 32%, mechanical movement 30%, pumps 19%, and fans 19%. It follows from these values that pumps are responsible for about 8—9% of the global consumption of electric energy.

According to Ref. [21], *in the EU* electric motors converted 1300 TWh of electricity into mechanical energy in 2012, corresponding to 520 Mt of CO_2 emissions. This value is expected to increase to around 1500 TWh in 2020 and 1800 TWh in 2030.

In the industrial sector in the EU, EMDSs are by far the most important electric energy consumers and use about 70% of the totally consumed electric energy, while in the tertiary sector, EMDSs use about one-third of the consumed electric energy [8]. In both sectors, EMDSs comprise compressors, refrigerator systems, pumps, ventilations, conveyors, and other equipment.

In the EU, the share of *pumps* in the annual consumption of electric energy by motor-driven systems was 21% in the industrial sector and 16% in the tertiary sector for the year 2000 [8].

The share of consumption of electric energy by motors in respect to their nominal power is described in Ref. [5] as follows:

Small-size electric motors with a nominal output power of less than 0.75 kW are the great majority and are applied primarily in the residential and commercial sectors. But these motors account for only about 9% of all electric energy consumed by motors.

About 68% of the electric energy consumed by electric motors is used by medium-size motors with a nominal output power of 0.75 to 375 kW. These are mostly AC induction motors with two to eight poles, but some are special motors (e.g., direct current, permanent magnet, switched reluctance, stepper, and servo motors). They are manufactured in large series according to standard specifications and can be ordered from catalogs. These motors account for about 10% of all motors.

Large electric motors with more than 375 kW nominal output power are usually high-voltage AC motors that are custom designed. They comprise just 0.03% of the electric motor stock but account for about 23% of all electric energy consumption by motors.

In the EU, the market shares of AC induction motors are 50−70% four-pole motors, 15−35% two-pole motors, and the remainder six- and eight-pole motors [8].

Electric motors used as pump drives are dominated by AC induction motors in the power class ≥ 0.75 kW. In the EU-25[1] market in the low-power range (<5 kW), the share of other motor types that have better efficiencies than AC induction motors shows a trend to increase on a low level. For example, in the EU-25, the share of permanent magnet motors was only 1.4% in 2002 [8].

Some typical functions of the motor-driven pumps are as follows [5]:

In the residential sector, the pumps serve for central heating systems, circulation of hot and cold water, and pressure boosting of water supply. In the commercial building sector, pumps are used for heating, ventilating, air conditioning, and water supply, including pressure boosting. In the agricultural sector, motor-driven pumps serve for irrigation. In industrial applications, which contribute the largest share of the electric energy consumed by motors (including also, e.g., air compression, conveyance, mechanical handling, and processing), pumping and pressurizing of water and other liquids is achieved by the use of electric motor driven pumps.

Generally, existing pumps fulfill their function on the basis of various physical principles of energy transfer to a liquid. These are

- rotodynamic pumps, which convert mechanical to fluid energy by the means of a rotating impeller
- positive displacement pumps, which convert mechanical to fluid energy by displacing geometrically the liquid from the low-pressure to the high-pressure side.

The rotodynamic pumps can be further subdivided into various types in respect to their geometry and constructional structure (radial, semi-axial, or axial impeller and casing geometry; single stage or multistage pumps; pumps with their own bearings or closed coupled to a motor; separate or integrated pump and motor casing; and others). Some special types of rotodynamic pumps are integrated with special types of electric motors. For example, submersible multistage pumps and most circulators are mechanically integrated with wet motors. In wet motors, in contrast to conventional electric motors, the motor rotor rotates in liquid and not in air.

The geometrical size of rotodynamic pumps and their nominal values of the mechanical or—in the case of integrated pump units—electric power

---

1. EU-25 refers to all EU-25 countries except Malta and includes Switzerland.

input cover a very wide range. The nominal power input values range from less than 100 W (e.g., in the case of circulators for buildings, see below) to several megawatts (e.g., for large boiler feed pumps).

The positive displacement pumps comprise two subcategories: pumps of the reciprocating or of the rotating type. Piston pumps, diaphragm pumps, and plunger pumps represent the first subcategory. Gear pumps, screw spindle pumps, and progressing cavity pumps are examples of the second subcategory.

Regarding their contribution to the total electric energy consumed by motor-driven pumps, rotodynamic pumps are by far the most important. Therefore, the focus of this book is on rotodynamic pumps. Their cumulative effect on the current and future consumption of electric energy is determined by

● their numbers in use and in yearly production and/or replacement
● their individual nominal input power as well as their individual operating mode and its influence on their actual electric power input, averaged over the operating hours.

Rotodynamic pump types, which were identified as first product categories for regulatory measures by the European Commission (EC), are clean water pumps for applications in commercial buildings, drinking water pumping, food industry, agriculture, and circulators in buildings (see Chapter 2).

For the first category of pumps, the nominal mechanical power input is in the range from 0.75 to 150 kW.

For the installed stock of circulators, two subcategories and their corresponding typical values of nominal electric power input can be distinguished: For stand-alone circulators, which are separate from the boiler and are purchased as a separate product, typical values of the nominal electric power input are 65 W for applications in single houses and 450 W for applications in residential or commercial buildings. For boiler-integrated circulators, which are supplied to the user already integrated into a boiler, a typical value of the nominal electric power input is 90 W [7].

Concerning their contribution to the stock and to the market, the following numbers for the EU-25 were determined in studies mandated by the EU and reported in Refs [6,7]:

Number of pumps (types as defined in Ref. [6], see also Chapter 2):

Installed (estimated):	17.05 millions
Sales in 2007:	1.55 millions

Number of circulators (boiler-integrated and stand-alone)

Installed (estimated):	140.00 millions
Sales in 2005:	14.00 millions

In the same studies [6] and [7], the total consumption of electric energy in the EU-25 in 2005 is estimated to have been approximately

- 109 TWh for pumps (of the types defined in Ref. [6], see also Chapter 2)
- 50 TWh for circulators.

This corresponds to CO_2 emissions of

- 50 Mt by the use of pumps (of the types defined in Ref. [6], see also Chapter 2)
- 23 Mt by the use of circulators.

If no specific measures are taken, electricity consumption in 2020 is predicted to increase to

- 136 TWh for pumps (of the types defined in Ref. [6], see also Chapter 2)
- 55 TWh for circulators.

1.2 RELEVANT TECHNICAL FEATURES OF PUMPING SYSTEMS

From the facts and numbers given in Section 1.1.1, it is evident that technical systems with electric motor driven *pumps* (called in this context "pumping systems") contribute substantially to the global and especially to the European consumption of electric energy and indirectly to CO_2 emission.

In respect to the potential and to the technical and legislative measures that aim at improving the energy efficiency of *pumping system*s and thereby at saving electric energy, it is advisable to treat complete pumping systems as being composed of two main parts (see also Chapter 8):

1. The *pump unit* in the form of
   - a *single pump unit* as a combination of one pump and its electric drive
   - a *multiple pump unit* as a combination of at least two pumps and their corresponding drives closely assembled and operating together
     where the drive can be
   - either an electric motor fed directly from the electric grid
   - or an electric motor combined with and fed from a variable-speed drive (VSD)[2] in the form of a frequency converter. This combination of motor and VSD is called in this context Power Drive System (PDS).
2. The *hydraulic installation*, which consists of pipes, valves, and other equipment (e.g., heat exchangers, liquid reservoirs, and others) and is supplied with fluid energy by the pump unit

---

2. More familiar in the field of electric equipment is the denomination Complete Drive Module (CDM); see also Section 5.2.

The hydraulic installation can be of the open or closed type (see Chapter 8) and is the "final consumer" of energy in the form of fluid energy.

In the process of designing a hydraulic installation and dimensioning its components—which is normally not the responsibility of the manufacturer or supplier of the pump unit(s)—for the demand of flow rate, usually a design value of the flow rate Q_{design} is specified. The design value of demanded flow rate for an individual pumping system is specified, for example, by an expected withdrawal of water from water supply systems or by the heat to be transferred in heating systems in their design condition. For this value Q_{design} (usually called $Q_{100\%}$) the corresponding design value of the required head H_{design} (usually called $H_{100\%}$) is determined, mostly by calculation on the basis of known or assumed hydraulic loss or resistance coefficients of the individual components of the hydraulic installation.

Note: The head H, usually given in the unit m (meter), is a physical quantity that is equivalent to a difference of fluid energy.

However, nearly all pumping systems are not operated permanently at their design condition. The demand of flow rate Q_{demand} varies during the running time of the pumping system within a range

$$Q_{demand,min} \leq Q_{demand} \leq Q_{demand,max}.$$

Depending on the application, this range can be rather small or quite broad. Applications that are characterized by a rather small range of demanded flow rate can be categorized as "(nearly) constant flow systems." Applications that are characterized by a quite broad range of demanded flow rate can be categorized as "(widely) variable flow systems." In both categories, the maximum demanded flow rate $Q_{demand,max}$ is usually equal to (or only slightly greater than) the design flow rate. In installations of the variable flow category, the flow rate $Q_{demand,max}$—which is for this category usually also the design flow rate $Q_{100\%}$—is only demanded at a relatively short part of the total running time. In many applications of the variable flow category, the minimum flow rate $Q_{demand,min}$ that is demanded during the running time of the pumping system is considerably lower than the design value $Q_{100\%}$ and is, for example, in pressure boosting systems for buildings only a few percentages of $Q_{100\%}$ (see Chapter 8).

The frequency distribution of the demanded flow rate over the running time of a pumping system is called *flow-time profile* in this book. This flow-time profile is usually described by dividing the range of demanded flow rate into a number of equal intervals—represented by the mean value of Q_{demand} within the respective interval—and stating the fraction of the total operating time corresponding to the respective intervals:

$$\frac{\Delta t}{t_{tot}} = f(Q_{demand}) \tag{1.1}$$

or more often in the dimensionless form

$$\frac{\Delta t}{t_{\text{tot}}} = f\left(\frac{Q_{\text{demand}}}{Q_{100\%}}\right) \tag{1.2}$$

For example, by extensive studies and measurement in the field of building service engineering in Europe, characteristic and representative flow-time profiles were found for the applications "heating" and "pressure boosting in buildings"; see Refs [9,10], and Chapter 8.

With the variation of the demanded flow rate Q_{demand} during operation of the pumping system, also the demanded head H_{demand} of the installation varies in dependence of Q_{demand}. For the dependence $H_{\text{demand}}(Q_{\text{demand}})$ different cases have to be distinguished.

1. Pumping Systems Without Feedback and Control of Pressure

In pumping systems without feedback and control of pressure of the hydraulically *closed type*, the minimum demand of head is exclusively determined by the sum of inherent hydraulic (friction) losses in the piping and all other loop components. In good approximation, the demanded head H_{demand} attributed to the hydraulic losses, called H_{loss}, is proportional to the square of the demanded flow rate Q_{demand}.

$$H_{\text{demand}} = H_{\text{loss}} = k_{\text{loss,tot}} \cdot Q_{\text{demand}}^2 \tag{1.3}$$

By relating the variable values of demanded flow rate and demanded head to their design ($=100\%$) values, Eq. (1.3) can be expressed in the dimensionless form:

$$\frac{H_{\text{demand}}}{H_{100\%}} = k_{\text{loss,tot}}^* \cdot \left(\frac{Q_{\text{demand}}}{Q_{100\%}}\right)^2 \tag{1.4}$$

The coefficients $k_{\text{loss,tot}}$ and $k^*_{\text{loss,tot}}$, respectively, result from the totality of hydraulic losses in the hydraulic installation.

In pumping systems without feedback and control of pressure of the hydraulically *open type*, the demanded head comprises additionally to the losses a (static) part H_{static}, which is independent from the demanded flow rate Q_{demand}. It results from a geodetic height difference by which the pumped liquid has to be lifted. In this case, the demanded pump head is

$$H_{\text{demand}} = H_{\text{static}} + k_{\text{loss,tot}} \cdot Q_{\text{demand}}^2 \tag{1.5}$$

or in the dimensionless form

$$\frac{H_{\text{demand}}}{H_{100\%}} = \frac{H_{\text{static}}}{H_{100\%}} + k_{\text{loss,tot}}^* \cdot \left(\frac{Q_{\text{demand}}}{Q_{100\%}}\right)^2 \tag{1.6}$$

2. Pumping Systems with Feedback and Control of Pressure

In various applications of pumping systems, especially in building services (heating, cooling, water supply) the hydraulic installations are of the branched type. In these cases, there usually exist additional requirements in respect to minimum values of pressure or pressure difference at points of the installation that can also be located rather far from the pump. For example, in pressure boosting systems for tall buildings, for any total amount of water taken from the system at the same time the water pressure shall have a sufficiently high value at any tapping point. Therefore, a pressure control is implemented in the pumping system, which serves to fulfill such a pressure requirement. This requirement is mathematically described in the form of so-called *pressure control curves* (or functions) for the demanded head $H_{demand} = f(Q_{demand})$. These functions can be adapted to a special type of application or even to an individual application, but usually functions that are representative for certain types of applications are realized in practice. They are usually described in dimensionless form by relating the variable values of demanded flow rate and demanded head to their design ($=100\%$) values and are often defined as linear functions of the form

$$\frac{H_{demand}}{H_{100\%}} = \frac{H_0}{H_{100\%}} + \text{const} \cdot \left(\frac{Q_{demand}}{Q_{100\%}}\right) \tag{1.7}$$

For example, for the application field "pressure boosting in buildings" in Europe, a characteristic and representative function was deduced; see Ref. [10]. Graphical representations of relations according to Eqs (1.4), (1.6), and (1.7) are shown in Figure 1.1.

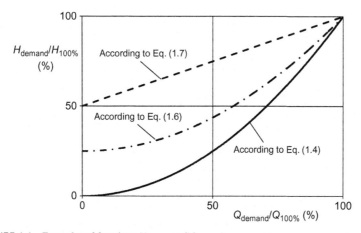

FIGURE 1.1   Examples of functions $H_{demand} = f(Q_{demand})$.

1.2.1   The Demanded Hydraulic Power

The demanded values of flow rate and head determine the demand of hydraulic power

$$P_{\text{hyd, demand}} = \rho \cdot g \cdot Q_{\text{demand}} \cdot H_{\text{demand}} \tag{1.8}$$

which—for its part—is determined by (and proportional to) the demand of flow rate Q_{demand} and corresponding pump head H_{demand}. In Eq. (1.8), ρ is the fluid density and $g = 9.81$ m/s^2 is the gravitational constant.

Inserting the values $Q_{100\%}$ and $H_{100\%}$ into Eq. (1.8) leads to the design ($=100\%$) value of the demand of hydraulic power $P_{\text{hyd,100\%}}$ corresponding to the design operating condition:

$$P_{\text{hyd, 100\%}} = \rho \cdot g \cdot Q_{100\%} \cdot H_{100\%} \tag{1.9}$$

The values $Q_{100\%}$, $H_{100\%}$, and $P_{\text{hyd,100\%}}$ at the design operating condition of the installation are usually taken as the basis for the design of the complete pumping system, including the specification and selection of the pump unit. Of course, with the variation of the demanded flow rate according to the respective flow-time profile, the demanded hydraulic power also varies (according to Eq. (1.8)) during operation of the pumping system. In dimensionless form, the relation for the demand of hydraulic power can be written:

$$\frac{P_{\text{hyd, demand}}}{P_{\text{hyd, 100\%}}} = \frac{Q_{\text{demand}}}{Q_{100\%}} \cdot \frac{H_{\text{demand}}}{H_{100\%}} \tag{1.10}$$

where the ratio $H_{\text{demand}}/H_{100\%}$ is a function depending on the type of the pumping systems, for example described by Eqs (1.4), (1.6), or (1.7), respectively. The corresponding relations for the demanded hydraulic power are shown graphically in Figure 1.2.

1.2.2   Pump Units

The purpose of pump units is to convert electric energy ($=$input) into fluid energy ($=$output). This conversion is effected in two steps.

In the electric drive of a pump unit (consisting of only motor(s) or motor(s) combined with CDM(s)), the electric energy ($=$input) is converted into mechanical energy ($=$output) which is transmitted to the pump(s) by a rotating shaft. The efficiency of an electric drive is the ratio of the mechanical output power P_{mech} to the electric input power P_1.

$$\eta_{\text{drive}} = \frac{P_{\text{mech}}}{P_1} \tag{1.11}$$

with

$$P_{\text{mech}} = \omega \cdot T \tag{1.12}$$

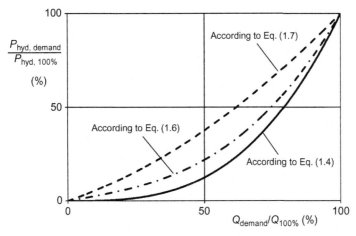

FIGURE 1.2 Examples of functions $P_{\text{hyd,demand}} = f(Q_{\text{demand}})$.

as the product of the angular velocity ω and the shaft torque T. The angular velocity ω (in the unit 1/s) results from the rotational speed n (in the usual unit 1/min) by the relation

$$\omega = \frac{2\pi \cdot n}{60} \tag{1.13}$$

The electric components (motor, CDM, or complete PDS) are usually classified (see Chapter 2) and selected in respect to their nominal efficiency—(i.e., the efficiency at their nominal or "rated" operating condition—mechanical power, shaft rotational speed). But it is important to realize that in pumping units (and in the majority of all other motor driven systems) they are mostly operated at variable load conditions apart from the nominal operating point. According to Ref. [8], the so-called load factor (which is the ratio of the actual average load to the rated mechanical output power) of electric motors for many fields of application (including pumping systems) and for motor power classes from 0.75 to 500 kW is typically about 0.6. While the efficiency of standard AC induction motors varies only slightly for operating points in the range from 75% to 100% of nominal load (=power output), it normally drops sharply for operating points below 50% of nominal load (see Section 5.1). This effect is more pronounced for small motors.

In the pump, the mechanical energy (=input) is converted into hydraulic energy (=output). The efficiency of a pump is the ratio of the hydraulic power P_{hyd} to the mechanical input power P_{mech}.

$$\eta_{\text{pump}} = \frac{P_{\text{hyd}}}{P_{\text{mech}}} \tag{1.14}$$

wherein the useable output power of the pump is

$$P_{\text{hyd}} = \rho \cdot g \cdot Q_{\text{pump}} \cdot H_{\text{pump}} \tag{1.15}$$

In Eq. (1.15) Q_{pump} is the actual flow rate delivered by the pump and H_{pump} is the actual head generated by the pump.

For a constant speed n the pump efficiency is strongly dependent on the delivered flow rate Q_{pump}. It shows a distinct maximum at the operating point of best efficiency (BEP) and the corresponding pump flow rate Q_{BEP} and decreases continually with increasing deviation of Q_{pump} from Q_{BEP}. The shape of the curve $\eta_{pump} = f(Q_{pump})$ is—in a fundamental manner—dependent on the pump type and size and particularly on a number that is called "specific speed" n_s and which characterizes the form of the pump impeller (from strictly radial to axial); see also Chapter 4.

Combining both steps of energy conversion in a pump unit and their corresponding efficiencies, the efficiency of the complete pump unit results as

$$\eta_{unit} = \eta_{pump} \cdot \eta_{drive} = \frac{P_{hyd}}{P_1} = \frac{\rho \cdot g \cdot Q_{pump} \cdot H_{pump}}{P_1} \qquad (1.16)$$

1.2.3   Methods of Varying and Adjusting the Flow Rate

To vary the flow rate in pumping systems and adjust it to the value demanded at a time, the main methods in use are[3]:

1. Variation of the total hydraulic resistance of the installation by variable throttling, for example by the means of control valves
2. Variation of the pump speed
3. Switching on or off individual pumps that are assembled in parallel mode (only possible in multiple pump units)
   3.1. without speed variation of one or more of the individual pumps
   3.2. in combination with speed variation of one or more of the individual pumps.

For all these methods, the pump flow rate Q_{pump} is equal to the demanded flow rate Q_{demand}.

If the variation and adjustment of the flow rate in a pumping system are realized by varying the pump speed (previously mentioned methods 2 and 3.2), the pump head H_{pump} is equal to the demanded head H_{demand} for any operating condition. See Figure 1.3 for an exemplary pump $Q-H$ curve and a function of demanded head according to Eq. (1.7).

Thereby, the hydraulic power generated by the pump is equal to the hydraulic power demanded by the installation.

On the other hand, if the variation and adjustment of the flow rate in a pumping system is realized by variable throttling, the pump head H_{pump} has to cover the net head $H_{demand,net}$ demanded by the hydraulic installation plus the loss of

---

3. Other existing methods as, for example, a bypass line with adjustable hydraulic resistance, are less relevant in the context of this book.

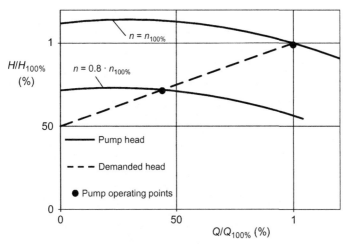

FIGURE 1.3   Adjustment of the flow rate by variable pump speed (exemplary).

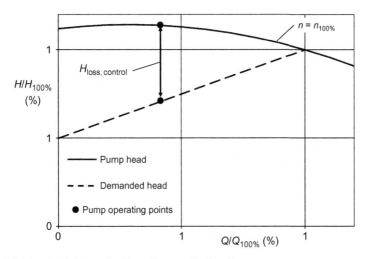

FIGURE 1.4 $Q-H$ diagram for flow adjustment by throttling.

head $H_{\text{loss,control}}$ which results from throttling the flow in the control valve. Because of the typical shape of the pump characteristic $H_{\text{pump}} = f(Q_{\text{pump}})$—also called $Q-H$ curve—the loss of head resulting from the control valve increases with decreasing flow rate; see Figure 1.4 for an exemplary pump $Q-H$ curve and a function of demanded head according to Eq. (1.7).

In the control valve, the part

$$P_{\text{loss,control}} = \rho \cdot g \cdot H_{\text{loss,control}} \cdot Q \tag{1.17}$$

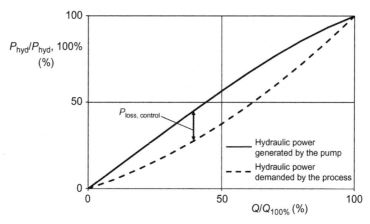

FIGURE 1.5   Loss of fluid power by throttling.

of the hydraulic power generated by the pump is dissipated. The dependence of this power loss on the flow rate is shown in Figure 1.5 for an exemplary pump $Q-H$ curve and a function of demanded head according to Eq. (1.7).

1.2.4   Relation of Demand and Actual Consumption of Energy in Pumping Systems

For the complete conversion of electric energy supplied by the electric grid into the demanded hydraulic energy of the respective application, the total efficiency of energy conversion is

$$\eta_{\text{tot,conversion}} = \frac{P_{\text{hyd, demand}}}{P_1} = \frac{P_{\text{hyd, demand}}}{P_{\text{hyd,pump}}} \cdot \frac{P_{\text{hyd, pump}}}{P_1} = \frac{H_{\text{demand}}}{H_{\text{pump}}} \cdot \eta_{\text{pump}} \cdot \eta_{\text{drive}}$$

(1.18)

All factors on the very right side of this equation depend on the flow rate Q, except in the case of flow adjustment by variable speed, where the ratio $H_{\text{demand}}/H_{\text{pump}}$ is generally equal to 1.

Taking into account the flow-time profile of operation, the hydraulic energy E which is actually demanded by a pump system sums up to

$$E_{\text{demand}} = t_{\text{tot}} \cdot \sum_{i=1}^{i=N} \left(\frac{\Delta t}{t_{\text{tot}}}\right)_i \cdot (P_{\text{hyd,demand}})_i$$

(1.19)

In this equation, t_{tot} is the number of operating hours in a certain period of operation relevant for assessing the energy efficiency, i is the consecutive number, and N is the total number, respectively, of the flow rate intervals of the flow-time profile.

On the other hand, the consumption of electric energy $E_{consumption}$ by the same pump system in the same period of operation is

$$E_{consumption} = t_{tot} \cdot \sum_{i=1}^{i=N} \left(\frac{\Delta t}{t_{tot}}\right)_i \cdot (P_1)_i \tag{1.20}$$

Combining Eqs (1.18) and (1.20) yields

$$E_{consumption} = t_{tot} \cdot \sum_{i=1}^{i=N} \left(\frac{\Delta t}{t_{tot}}\right)_i \cdot \left(\frac{\rho \cdot g \cdot Q_{demand} \cdot H_{demand}}{\eta_{pump} \cdot \eta_{drive}} \cdot \frac{H_{pump}}{H_{demand}}\right)_i \tag{1.21}$$

In this equation

- the pump efficiency depends on the pump flow rate Q_{pump} (which is equal to the demanded flow rate) and in the case of variable speed operation on the rotational speed n (see Sections 4.2 and 5.2)
- in the case of a grid-fed motor the drive efficiency η_{drive} is equal to the motor efficiency η_{motor} and depends on the mechanical power P_{mech} (see Section 5.1)
- in the case of a PDS (motor + CDM) the drive efficiency η_{drive} is equal to the efficiency of the PDS η_{PDS} and depends on the shaft torque T and on the rotational speed n (see Section 5.2).

The mechanical power, the shaft torque, and—in the case of flow rate adjustment by throttling—also the ratio H_{pump}/H_{demand} depend on the demanded values of flow rate and head corresponding to the respective operating condition of the pumping system.

It follows therefore that the consumption of electric energy by a pumping system over a certain period of operation is

- not only and simply determined by the efficiency values of its components (pump, motor, and, if available, CDM) at their respective nominal operating condition
- but is also (and even considerably more) determined by
  - the range of demanded flow rate and the flow-time profile of the respective application
  - the relation $H_{demand} = f(Q_{demand})$ of the respective application
  - the efficiencies of the components within the whole range of corresponding operating conditions (which are particularly part load operating conditions).

1.3   SAVING ELECTRIC ENERGY USED FOR PUMPING SYSTEMS

Generally, saving energy and the reduction of CO_2 emission (in a global, European, or national frame) can be reached by various attempts and measures concerning the generation as well as the consumption of electric energy.

In respect to the goal of *saving consumed energy*, the major EU policy package, called the "climate and energy package," aims at a 20% reduction in primary energy use compared with projected levels, to be achieved by *improving energy efficiency* of energy consuming users and equipment.

1.3.1 Technical Measures to Improve Energy Efficiency of Pumping Systems

As explained in Section 1.1.2, EMDSs contribute by a remarkable amount to the total consumption of electric energy. Since no reduction of their total number in use can be expected in the future, the only way to reduce their cumulative energy consumption consists of improving their individual energy efficiency and/or efficient operation. For this improvement, a considerable potential exists by technical measures (see also Chapters 4 and 5). The possible improvement for industrial motor-driven systems in the EU is estimated to 25% by [4] and to 20–30% by [8].

According to Refs [4,8], the implementation of minimum efficiency levels for motors sold in the EU (see Section 2.2) from 2011 on would result in cumulative savings of 72–92 TWh by 2020.

In both reports [4,8], three major fundamental contributions to practically achieve the improvement are indicated:

- To increase the use of more efficient motors
- To increase the use of VSDs where they are appropriate
- To optimize the complete system, including correct sizing of all parts of the system.

In respect to *pump systems* and their special features and relations described in Section 1.2, the technical measures to save electric energy can be pointed out and completed here in more detail.

1.3.2 Use of More Efficient Components of Pump Units

As for the totality of EMDSs (see Section 1.1.2), also for pump units, AC induction motors are by far the most used motor type. They are standard products, cheap and well suited to continuously operate pumps at fixed speed when fed directly from the grid or at variable speed in combination with a CDM.

The energy efficiency of grid-fed AC induction motors is presently classified by its efficiency at the nominal motor operating condition (=mechanical power output) and described by the classes IEx (with x

being an integer number ≥ 0); see Chapter 2. The use of more efficient motors can be technically realized by

- motors of a better efficiency class
- use of other motor types (e.g., electronically commutated or permanent magnet motors as well as switched reluctance motors) that have better efficiencies than AC induction motors.

In respect to the effect of using motors of a better IE class, it has to be pointed out that there is a difference of only a few percentage points in energy efficiency between average motors and the most efficient motors on the market, and that there is only a very limited potential for further efficiency improvement in the state-of-the-art electric motors on the market today. According to Ref. [5], the potential to improve the energy efficiency of AC induction motors lies mainly in

- smaller motors (<10 kW), because their efficiency difference from a worse IE class to the better ones is larger than for bigger motors
- increased efficiency at part load operation.

Regarding pump units, the efficiency of the pumps is equally important as that of the drives. Therefore, the use of more efficient pump units must not be related to the motors alone but must also concern the pumps.

Furthermore, it is not sufficient to select the components of pump units exclusively in respect to their efficiency values at *nominal conditions*. In fact, in most applications it is important to select pump units and/or their components in respect to lower cumulative demand, which—on its part—is additionally and even more strongly influenced by

- the flow-time profile and the relation $H_{\text{demand}} = f(Q_{\text{demand}})$ of the respective (type of) application
- the efficiency of the components and—most important—of the pump unit as a whole in the whole range of flow rates during operation (which is typically an operation at *preferably part load conditions* of the unit and its components).

1.3.3 Use of Variable Speed Driven Pump Units

For many motor-driven systems, including pumping systems, the use of motors in combination with a frequency converter (instead of motors fed directly from the electric grid and operated at fixed speed) can principally serve to improve energy-efficient operation and to reduce consumption of electric energy.

The two parts of a PDS, which is the combination of an electric motor and a CDM, can be

- either produced and sold as a package, preferably as an integrated product
- or manufactured and sold by different manufacturers and assembled together with the driven machine after purchase as a nonintegrated product.

According to Ref. [8], recent industry-based estimates in Germany show that 30% of electric motors are sold together with a CDM. Small pumps are also increasingly sold in integrated packages that include a CDM.

However, the usefulness and energy-saving effect of VSDs for pump units depend considerably on the type of application and corresponding mode of operation, especially on the flow-time profile. In the case of a nearly constant flow type of operation, normally a VSD brings no (or not economically significant) advantage in respect to energy consumption compared to a fixed-speed drive because of the additional losses in the CDM *and* in the converter fed motor (see also Section 5.2). Therefore, a remarkable improvement of the energy efficiency by VSDs for pump units will mainly result for applications with the operational mode "(widely) variable flow." But even for this operational mode, in the case of multiple pump units (e.g., pressure boosting sets), there exists the possibility to vary the flow rate by switching on or off individual pumps, which are arranged in parallel in combination with a pressure control device. For such pump units, the possible improvement by variable-speed driving depends on their hydraulic design (number and size of individual pumps) and their control as well as on the flow-time profile and pressure control curve. It may be sufficient and economically preferable to combine only one of the individual pumps with a VSD to already reach the most significant effect of variable speed driving on the cumulative energy consumption of the whole unit.

Furthermore, the statement that for pumps operated at variable speed, the mechanical power demanded from the drive is proportional to the third power of the pump (and motor) speed (which can be found in many studies and corresponding reports or other publications), is not generally true. According to the facts and features explained in Section 1.2, this relation between rotational speed and mechanical power is only valid (at least in good approximation) for pumping systems where the head demanded by the installation and generated by the pump obeys a relation according to Eq. (1.3) (i.e., is proportional to the square of the flow rate). In all other cases of relations between demanded head and flow rate, the variation of the mechanical power with the rotational speed obeys a more or less different relation. Nevertheless, VSDs will always lead to a significant reduction of energy consumption for operational modes of the type "(widely) variable flow."

Of course, also for pump units with VSDs, the actual (cumulative) consumption of electric energy from the grid is additionally dependent on the flow-time profile and on the unit efficiency within the range of demanded flow rate (see Eq. (1.21)). The drive efficiency, in the case of variable-speed driven motors the efficiency of the PDS, determines—together with the pump efficiency—the unit efficiency. Therefore, higher PDS efficiencies (in the range of operating points, defined as combinations of motor shaft torque and rotational speed) contribute to the unit energy efficiency. According to Eq. (1.16), the unit efficiency is directly proportional to the efficiency of the PDS.

The energy efficiency of a PDS is currently classified by its efficiency at the *nominal* motor operating condition (=mechanical power output) and described by the classes IESx (with x being an integer number ≥ 0); see Ref. [15] and Section 2.1. The use of more efficient PDSs can be technically realized by

- motors of a better efficiency class
- frequency converters with better efficiency $\eta_{converter}$ (efficiency classes for converters are also defined in Ref. [15])
- optimization of the PDS in respect to the interactions of motor, converter, and control software (see Chapter 5), especially in the case of integrated PDSs.

1.3.4   Proper Design, Sizing, and Dimensioning of the Complete Pumping System

Concerning measures in the frame of designing, sizing, and dimensioning the components of pumping systems, which aim at improving their energy efficiency and/or energy-efficient operation, the respective responsibilities must be emphasized.

The design and dimensioning of the hydraulic installation has direct effect on flow velocities and friction losses in the piping and other components of the installation and thereby on the demanded head. But this work is most often carried out by the end user, by an engineering company, or by others and is *not in the responsibility of the manufacturer or supplier of the pump unit*. This work includes also the specification of the design values of the complete system and thereby the specification data for the selection of the pump unit(s). Measures to improve the energy efficiency of the complete system are

- proper dimensioning and choice of all components of the hydraulic installation regarding the hydraulic losses attributed to them with the aim to deliver the required hydraulic energy to the process with minimal energy losses
- avoiding or rationally minimizing "safety margins" in the specification of design values that shall cover uncertainties in predicting the demanded flow rate and head and lead to an oversizing of the pump and corresponding pump drive.

On the other hand, the *manufacturer or supplier of the pump unit and/or its components* can—in the frame of its responsibility—also contribute to avoid unnecessary oversizing and worse energy efficiency caused thereby.

In the case where the pump unit is sold as a prefabricated product by a pump manufacturer, it should

- consist of optimally matched components (pump, motor, or PDS) regarding their nominal values and part-load efficiencies
- be selected for the respective application with best possible matching of the required and delivered hydraulic data (flow rate, head).

In other cases, the components of a pump unit (pump, motor, and, if available, CDM or integrated PDS) are purchased from different manufacturers and assembled to a final product by another company. In these cases,

- the pump should be selected as in the case mentioned previously
- the motor (without or with a CDM) should be sized correctly regarding the mechanical power required by the pump.

As mentioned in Ref. [5], there is a tendency in practice to oversize motors based on a misguided belief that larger motors will operate more reliably for a given application. But new high-efficiency motors can be sized with less safety. Typical oversized motors or PDSs have efficiency disadvantages. Because of the severe decrease in efficiency at low load, especially for small motors, even the benefits of using a more efficient motor can be lost if the load factor (as defined in Section 1.2.2) is very low [8].

Of course, the CDM should be matched with the motor without unnecessary oversizing in case a motor and CDM are selected and purchased separately.

1.3.5   Estimations of Achievable Improvements and Savings

In various reports on studies, values of achievable improvements of efficiencies were given for certain cases and based on certain assumptions (concerning flow-time profile, type of hydraulic installation, pump types, and measures put into practice for improvement).

In Ref. [8], an example (taken from a former study [40]) serves to illustrate the cumulative energy-saving effects of

- use of a more efficient pump
- use of a more efficient motor
- use of a variable speed instead of variable throttling for flow adjustment
- slightly reduced pipe friction losses

which in sum lead to an improvement of "system efficiency" (=ratio of hydraulic power demanded by the process to the electric power input) from 31% to 72% for an operation point at 60% of the 100% value of flow rate.

In Ref. [5] it is stated (and illustrated by values for the particular measures) that a systematic integration and optimization of all mechanical and electrical components in a complete pumping or ventilation system can bring a benefit of improving the (system) energy efficiency from 42% to 63%.

Also in Ref. [5], an exemplary study served to demonstrate the possible savings of energy consumed by pumping systems if all available state of-the-art technical measures

- reduction of piping losses
- use of highly efficient permanent magnet motors with a CDM
- and correctly sized pumps

are applied. According to this study, savings of 80–90% can be achieved in "heating-system circulator pumps" and of 40–75% in "industrial size pumps."

Furthermore, in Ref. [8] a detailed analysis is carried out for pumping systems with a function $H_{demand} = f(Q_{demand})$ according to Eq. (1.3) (hydraulic installation of the closed type without pressure control) and an assumed flow-time profile according to Table 1.2.

In this calculation, the two cases of flow adjustment by throttling and by the use of a VSD are compared in respect to the consumption of electric energy for various motor sizes (1.1, 11, and 110 kW) and various numbers of annual pump operating hours. As a result, for an exemplary annual operating time of 4000 h and for the whole investigated range of motor sizes, the savings by the use of a speed variation for flow adjustment are about 36%.

The preparatory study [6], which concerns water pumps for commercial buildings, drinking water, agriculture, and the food industry, concludes that removing the worst 40% of pumps (regarding their pump efficiency) from the market would yield energy savings of 3.4 TWh/year by 2020, and the full impact of such an action is expected to be 4.8 TWh/year at the usage estimated for 2020.

In the preparatory study [7] that concerns circulators used in central heating systems, three different improved technologies are identified:

- Improved (standard) circulator
- Variable speed (AC induction motor)
- Variable speed (permanent magnet motor).

Energy savings of 13 TWh/year could be achieved by 2020 if the minimum energy performance of stand-alone circulators that can be sold is of the label A* (for the definition of labels for circulators that are now substituted by the Energy Efficiency Index (EEI), see Ref. [30] and Chapters 3 and 8).

TABLE 1.2 Flow-Time Profile

% of Flow Rate	% of Time
50	25
75	50
100	25

1.3.6 Measures to Promote Improved Energy Efficiency of Pump Units and Resulting Effects

As illustrated in Section 1.1.2, the energy efficiency of pump units is relevant for the total (global or European) consumption of electric energy. Therefore, it is an ecological and political goal to reduce energy consumed by the use of pump units.

In the EU, a legislative frame was established by a general directive concerning all energy using products (EuP) [11], also called the EcoDesign Directive. This directive allows the EC to develop measures to reduce the eco-impact of EuP within the EC. From this directive, implementing measures to promote energy savings in the use-phase and to reduce impacts on the environment in production and/or end of life for individual categories of products (to be done by the EC) are deduced. Besides many other products, electric motors and pumps are products which are also concerned; see Section 2.2. Pumps and electric motors (or more generally pumps and their electric drives) can be seen and treated as separate products if they are combined to pump units. This is called product approach; see Chapters 2, 3, and 7. But separate requirements on the energy efficiency of components of pump units will not utilize the full potential of saving energy consumed by pump units which can be achieved, for example, by variable-speed driving and/or "intelligent" control. Therefore, it is—in respect to saving energy—considerably more effective to focus on complete pump units as "extended products." This is called extended product approach; see Chapters 2, 3, and 8.

Defining and prescribing requirements on the energy efficiency of products in the frame of legislative measures requires practicable and meaningful methods for the description and assessment of the product energy efficiency and, if required, corresponding classification of the products. In contrast to some other products which are covered by the EcoDesign Directive, in the case of pump units neither the individual numerical efficiency values of the components (pump, electric motor, or PDS)—as defined in Eqs (1.11) and (1.14)—nor the individual numerical efficiency value of the complete unit—as defined in Eq. (1.16)—are sufficiently indicative to be directly taken for the assessment and classification.

This results from the fact that these component or unit efficiencies are not only determined by the product quality in respect to energy conversion but also by various other physically caused influences as, for example, the size, nominal rotational speed, nominal power, and some others (see Chapters 4 and 5). These physically caused influences cannot be overcome by better design and/or manufacturing quality, which is the responsibility of the product manufacturers and suppliers.

Therefore, for the assessment and classification of pump units and/or their components, efficiency classes (in the form of alphabetic characters without or with additional signs) or energy efficiency indicators (in the form of numerical values) are needed to express the energy efficiency and neutralize all influences on it besides the actual quality of the product in respect to energy conversion (= "efficiency related quality"). Examples are

- the efficiency levels or classes for electric motors, PDSs, and CDMs
- the Minimum Efficiency Index (MEI) for water pumps
- and the Energy Efficiency Index (EEI) for circulators and complete pump units (for details see Chapters 2, 7, and 8).

Requirements on minimum efficiency classes or minimum/maximum values of energy efficiency indicators with corresponding dates from when the products have to comply with are subjects of regulations elaborated by the EC and adopted by the European Parliament. From the date the requirements are valid, their fulfilment has to be indicated on the product or in the product documentation by the company that places it on the market and/or puts it into service. It is then an additional precondition for marking the product with the CE mark and for selling it within the EU. Products that miss complying with these requirements can ultimately be prohibited from being traded within the EC.

By these *legislative measures*

- manufacturers or suppliers of pump units are forced to eliminate from their product program those versions that don't comply with the requirements and to spent effort, time, and costs for improving or replacing them by the means of better design, manufacturing, and/or selection
- newly erected installations will be equipped only with pumps/pump units that comply with the required energy efficiency
- in existing installations, pumps, pump units, or their components will be progressively replaced by versions of better energy efficiency in the frame of replacement/refurbishing procedures.

Thereby, the legislative measures serve to achieve gradually the politically intended ecological goals.

However, consumption of electric energy by the use of electric motor driven pumps is also an economic aspect for their end users in the private, public, or industrial sector. In the great majority of applications, 90% and more of the life cycle costs of pump units result from the energy costs in the use-phase. Insofar, the legislative measures will also bring benefits

- for the end users who can operate their pumping system with reduced costs for energy and, thereby, reduced life cycle costs,
- but also for the manufacturers and/or suppliers of pump units that can— in the frame of market competition and their own marketing strategy— use, for example,

- products of better relation of the efficiency class or indicator to the price
- or products with "better than minimum required" efficiency class or indicator for increasing their market share.

On the other hand, in the frame of *standardization* not only classes or indicators of energy efficiency for the relevant categories of products should be defined and physically and technologically reasoned, but also (experimental and/or mathematical) methods to determine their individual values should be provided (see Chapters 2, 7, and 8).

Chapter 2

Standardization and Legislation Regarding the Energy Efficiency of Pump Units

This chapter reflects the situation in the fields of standardization and legislation concerning pump units and their components at the time this book was written (early 2015).

2.1 INTERNATIONAL STANDARDIZATION

In this chapter, some international standards that are of relevance for the assessment of the energy efficiency of pumps and pump units are discussed. Standards addressed here provide

- definitions of classes or indicators that are used to classify or quantify the energy efficiency of pump units or of their components
- and/or corresponding methods of measurements and calculations.

2.1.1 Standards Concerning Electric Components of Pump Units

The Standard **IEC 60034-30** of 2008 [11] (also adopted as European Standard **EN 60034-30**) defines efficiency classes by the corresponding minimum efficiency values for AC induction motors in a certain range of performance properties.

The standard covers single-speed, three-phase 50 or 60 Hz cage induction motors that are intended for direct on-line connection (=line-fed) and

- have a rated voltage U_N up to 1000 V and a mechanical power $0.75 \leq P_N \leq 375$ kW
- have up to six poles
- are rated for certain operating conditions defined in the standard.

Assessing the Energy Efficiency of Pumps and Pump Units.
DOI: http://dx.doi.org/10.1016/B978-0-08-100597-2.00002-1
© 2015 Elsevier Ltd. All rights reserved.

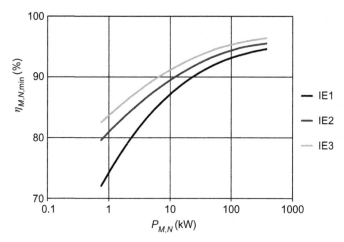

FIGURE 2.1   Minimum efficiency $\eta_{M,N,\mathrm{min}}$ of electric four-pole 50 Hz motors in dependence of the nominal motor power $P_{M,N}$ according to Ref. [11].

Explicitly excluded are (besides others)

- motors specifically made for converter operation with increased insulation
- motors completely integrated into a machine (pump,[1] fan, compressor, ...) which cannot be separated from the machine.

The standard defines three levels[2] of energy efficiency:

- IE1—standard efficiency
- IE2—high efficiency
- IE3—premium efficiency.

For each efficiency level (=class), pole number and grid frequency minimum values of motor efficiency are defined that have to be achieved or exceeded for motor operation *at nominal load* (i.e., with rated (=nominal) mechanical power). The values in dependence of efficiency level, pole number, grid frequency, and rated (=nominal) mechanical power are given as mathematical equations and additionally in tables for the standardized pump sizes. An exemplary graphical representation for four-pole 50 Hz motors is shown in Figure 2.1.

---

1. For example, wet motors integrated in circulators.
2. A fourth level, IE4 (Super Premium efficiency), was also introduced but not defined yet because of missing sufficient market and technological information for a standardization. This class is projected to have roughly 15–20% lower losses than the class IE3.

In respect to the experimental determination and documentation of the motor efficiency the standard prescribes:

- Efficiency and losses shall be tested in accordance with IEC 60034-2-1 [12]. For energy efficiency levels IE2 and IE3, only methods associated with low uncertainty shall be acceptable.
- The rated efficiency and the efficiency class shall be durably marked on the rating plate.

Presently, the standard is under revision. The already existing draft standard of March 2014 is now divided into two parts:

- Part 1—Efficiency classes of line-operated AC motors (IE code)
- Part 2—Efficiency classes of variable speed AC motors (IE code).

Part 2 is still under development.

According to Ref. [21], main changes in the draft of the revised version **IEC 60034-30-1** (March 2014) compared to the version of 2008 are as follows:

All technical constructions of electric motors are covered as long as they are rated for direct on-line operation. Thereby, for example, line-start permanent magnet motors are included.

The draft standard covers single-speed electric motors that are intended for direct on-line connection within the extended range of rated power $0.12 \leq P_N \leq 1000$ kW and with up to eight poles and are rated for certain operating conditions defined in the standard.

Explicitly excluded from the draft standard (besides others) are:

- motors completely integrated into a machine (for example pump, fan, and compressor) that cannot be practically tested separately from the machine
- submersible motors specifically designed to operate wholly immersed in a liquid
- motors with integrated frequency-converters (compact drives) when the motor cannot be tested separately from the converter.

In addition to the efficiency classes IE1, IE2, and IE3, now the IE4 efficiency level is included in the draft standard. An IE5 level is envisaged for a future revision with the goal of reducing the losses for the efficiency level IE5 by about 20% relative to IE4.

It is important to note that the efficiency levels in the power range $0.75 \leq P_N \leq 375$ kW already covered by the existing edition of the standard shall not be changed. Only below 0.75 kW and above 375 kW will the minimum required efficiency values be completed. The minimum required efficiency values for each class have to be achieved or exceeded regardless of the motor technology. From physical reasons, not all of the motor technologies covered by the revised standard will be capable of reaching the highest efficiency levels.

Currently, also Part 2 of the revised Standard **IEC 60034-30-2** is under consideration. It shall define efficiency levels (=classes) and corresponding minimum required efficiencies taking into account not only the nominal operating point but a number of load points at various pairs of rotational speed and shaft torque. It is expected that the harmonic losses (see Section 5.2) induced by the harmonics of the electric supply by the converter will contribute to the motor losses by an additional 15−20%.

The Standard **IEC 60034-2-1** [12] of 2007 describes test methods to determine the efficiency of electric motors. The efficiency of electric motors can be experimentally determined by two principally different test methods. The direct methods are input−output measurement methods. The input (=electric power) is determined by measuring the voltage and current of the electric supply, and the output (=mechanical power) is determined by measuring the shaft torque and rotational speed. The indirect methods are based on measuring the input and determining the total losses by summing up individual losses (see Section 5.1), which are individually measured (or calculated/estimated).

The standard describes in detail different test methods for the determination of motor efficiency with different levels of uncertainty. Three tables are presented with the preferred methods in combination with the required test facilities and associated levels of uncertainty.

The standard is currently being revised. The revision aims mainly at a better arrangement of the given information, at avoiding misinterpretations, and at some clarifications and refinements. Only very few technical modifications shall be introduced.

The Technical Specification **IEC/TS 60034-2-3** [13] of November 2013 focuses on test methods to determine the efficiency and losses of converter-fed AC induction motors. When fed by a frequency converter, in the motor occur—additionally to the fundamental losses—harmonic losses which result from the nonsinusoidal voltage and current waveforms generated by the converter. This Technical Specification[3] aims at providing methods to evaluate these harmonic losses.

Another IEC-document [14] should be mentioned that deals with the application of energy-efficient induction motors. It is intended to help responsible parties in the selection and use of adequate energy-efficient motors for various applications and motor-driven systems under consideration of various aspects.

---

3. Technical Specifications of IEC are published when the subject under question is still under development and/or when insufficient consensus for approval of an international standard is available.

The new European Standard **EN 50598** [15] is concerned with Ecodesign for power drive systems (PDSs), motor starters, power electronics, and their driven applications. It consists of three parts:

- Part 1: General requirements for setting energy efficiency standards for power-driven equipment using the extended product approach (EPA) and semi-analytic model (SAM)
- Part 2: Energy efficiency indicators for PDSs and motor starters
- Part 3: Environmental aspects and product declaration for PDSs and motor starters.

Part 1 and Part 2 are already published.

Part 1 is related to the fundamental aspects of the EPA.

Part 3 is not directly relevant for the aspects of assessment of energy efficiency presented in this book and is not addressed here.

In Part 2, various aspects of great importance for the electric components of pump units are dealt with.

As a basis for the classification of a PDS—(i.e., a combination of a motor and a complete drive module (CDM)) in respect to their energy efficiency, two concepts are introduced in the standard:

- The concept of the Reference PDS (RPDS)
- The concept of SAMs.[4]

It should be noted that the concept and the denomination of SAMs were originally proposed in the frame of the work of Europump (see Chapters 3 and 8) on the EPA for pump units.

The standardized RPDS serves to assess the energy consumption of an average technology PDS, consisting of a reference converter fed motor (RM) and a reference CDM (RCDM). The RPDS defined in the standard makes it possible to classify an actual PDS against the RPDS. For these classifications, limit values are set as described below.

For the determination of the losses and the efficiency of a RPDS having a certain rated (=nominal) power output, a calculation method is described in the form of mathematical models for losses that occur in the components (RM and RCDM) of the RPDS in dependence of the relative[5] shaft torque and the relative rotational speed. The losses are treated as "related losses" (i.e., they are quantified as dimensionless ratios of the actual losses to rated (=nominal) power values). In the case of the CDM, the losses depend on the (relative) output frequency and the (relative) output current as well as on the phase angle between fundamental output voltage and fundamental output

---

4. The use of semi-analytical models for the determination of the Energy Efficiency Index of pump units by calculations is also part of the methodology described in Chapter 8.

5. Relative means related to the nominal values, which are set to 100% values.

current. Therefore, the rotational speed of the motor has to be transferred into an output frequency of the CDM, and the motor torque has to be converted into an output current and a phase angle. The mathematical equations describing physical relations for the individual losses contain some parameters whose numerical values are taken from general experience. In addition to the calculation scheme and to the equations in the main body of the standard, some exemplary results are given in Annex A of the standard in the form of polynomial approximation equations and tables.

With the aid of the defined reference devices (RPDS, RM, and RCDM) energy efficiency classes are defined for actual devices (PDS, motor, and CDM) on the market. These classes are defined by the related losses of the actual devices compared to the corresponding related losses of the reference devices, both at the respective nominal power. For the converter-fed motor and for the CDM, the classes are denominated with IE plus the class number (beginning with zero). For the PDS, the classes are denominated with IES plus the class number (beginning with zero). The IE classes for the converter fed motors are under consideration to be defined in the future Standard IEC 60034-30-2 (see above). For the CDMs and the PDSs, in the present version of EN 50598-2 only the classes IE0 to IE2 and IES0 to IES2, respectively, are defined. The classes are

- IE0 and IES0, respectively, if the related losses are more than 25% higher than those of the reference devices
- IE1 and IES1, respectively, if the related losses are in the range of ±25% of those of the reference devices
- IE2 and IES2, respectively, if the related losses are more than 25% lower than those of the reference devices.

Besides the mathematical models for the calculation of losses, the standard describes test procedures for their experimental determination. Both methods can not only be applied to determine the losses at nominal operating condition (in the case of experiments called "type testing") but also at part load operating conditions. The latter is important for determining the Energy Efficiency Index (EEI) of complete units consisting of a PDS and a driven machine as, for example, of pump units; see Section 8.3. For this purpose, for each PDS (placed on the market as a complete product by one manufacturer) or for each motor and CDM that can be combined to a PDS eight load points as pairs of relative speed and relative shaft torque are defined for which the related losses shall be determined and documented by the respective manufacturer. The losses and the IE class of

- motors alone have to be determined (by tests or loss calculations) in combination with a RCDM
- converters alone have to be determined (by tests or loss calculations) in combination with a RM.

Also a method is described how to determine the losses and the IES class of a PDS by

- adding the losses of the CDM to the measured or calculated losses of the motor
- adding the losses of the motor to the measured or calculated losses of the CDM.

The application of the standard has the following consequences for the documentation by manufacturers or assemblers of the concerned products:

- A combination of a motor and a CDM placed on the market as one product (PDS) has to be marked with its IES class. The related losses at the requested load points also have to be indicated in the documentation.
- A CDM placed on the market as a product has to be marked with its IE class. The related losses corresponding to the requested load points have also to be indicated in the documentation.
- A motor placed on the market as a product has to be marked with its IE class for line-fed motors according to IEC 60034-30 (see above) until a standard for converter-fed motors will exist and define the corresponding classes. But the standard [15] gives sufficient information for the classification and loss determination of the combined drive system (motor + CDM).
- For a complete unit that consists of the three components (driven machine, motor, CDM) bought and assembled by a plant operator or machine builder, this company becomes responsible for the entire (assembled) product and its classification. However, according to information given in the standard [15], the IES class and the losses at the requested load points can be determined and documented for the PDS part.

2.1.2 Standards Concerning Pumps

2.1.2.1 Circulators

The European Standard **EN 16297** [16] of 2013 concerns the determination and assessment of the energy efficiency of circulators. Circulators are defined as glandless impeller pumps with or without pump housing designed for use in heating systems or cooling distribution systems. Pumps are denoted as glandless if the pump rotor of an electric motor is directly coupled to the impeller and immersed in the pumped medium.

The scope covers glandless circulators having a rated hydraulic output power of $1 \text{ W} \leq P_{\text{hyd},N} \leq 2.5 \text{ kW}$.

The document is split into three parts:

- Part 1: General requirements and procedures for testing and calculation of EEI
- Part 2: Calculation of EEI for stand-alone circulators
- Part 3: EEI for circulators integrated in products.

Stand-alone circulator means a circulator designed to operate independently from the product. Circulator integrated in a product means a circulator designed to operate dependently of the product. In this standard, the term "product" means an appliance that generates and/or transfers heat.

Part 1 defines

- the rated hydraulic power
- the corresponding values of $Q_{100\%}$ and $H_{100\%}$
- the reference pressure control curve
- the reference load profile (in this book called flow-time profile)
- the reference power P_{ref}

to be applied for the determination of the EEI. The measurements and calculations to be done for the determination of EEI are described in a general form. This includes the description of how the part load operating points have to be adjusted and measured and how the measured power input has to be compensated for possible deviations of the actual load points from the reference control curve.

The concept and methodology of the EEI are explained in more detail in Chapter 8.

The standard defines permissible distances between $H_{100\%}$ and curves published in the documentation (catalogs) according to the rated power input. Concerning accuracy of test equipment, it shall be in accordance with EN ISO 9906 [17], Grade 1, for measurements of flow, head, and power input. General requirements on test conditions are also stated.

The EEI shall be indicated in the form of two-digit decimals (e.g., EEI ≤ 0.21) together with the number of that part of the standard that had to be applied.

Parts 2 and 3 of the standard refer to Part 1 regarding the general aspects of testing and calculations and give specific information for the two different types of circulators, especially regarding the influence of specific speed on EEI for product-integrated circulators.

2.1.2.2   Water Pumps

In the state of finishing and final voting is the Standard **prEN 16480** [18] concerning the minimum required efficiency of water pumps. The scope of this standard will cover certain pump types used for clean water pumping:

- End suction and in-line pumps
- Vertical multistage pumps
- Submersible multistage pumps designed to operate in boreholes.

The pump types and corresponding technical properties and ranges of nominal operating conditions are specified in Annex A of the standard and correspond to the Ecodesign regulation EU (No.) 547/2012 [24]; see Section 2.2.3.

Besides some general definitions, the concept and definition of the Minimum Efficiency Index (MEI) is introduced. Mathematical equations are established that describe the correlation between MEI and the corresponding minimum efficiency values that are required at three relative[6] flow rates (75%, 100%, and 110% of the flow rate Q_{BEP} at the best efficiency point [BEP]). The first one is called "part load operating point," the third one "overload operating point." The inclusion of the minimum efficiency at the two other relative flow rates besides the nominal one (at BEP) in the definition of MEI is intended to take into account the energy efficiency of the pump when operating in a moderate range of flow rates around BEP. The correlation for the minimum efficiency at BEP contains a value C, which depends on MEI and additionally on the pump type and the nominal rotational speed. The minimum required efficiencies at the two other points are defined by multiplying the minimum required values at BEP with specified factors <1.

The concept and the methodology of the MEI are explained in more detail in Chapter 7.

Further, the standard contains descriptions of

- test procedures that have to be applied for the determination of MEI, including requirements concerning test conditions and allowable (random and device) uncertainties
- steps of evaluation of measurement results
- procedures for testing and/or evaluation of special pump types.

Because the pumps in scope are typically mass-produced pumps, the determination and indication of MEI generally refer to the mean value of a whole size. Information on the relevance of this fact and how it can be taken into account by the manufacturers is given in Annexes C and D of the standard (see also Chapters 6 and 7 of this book).

Concerning the evaluation of tests by the manufacturer, two procedures are described:

- Proving the MEI of a pump size in respect to compliance with requirements set by legislation (i.e., proving that $MEI_{actual} \geq MEI_{required}$)
- Determination of the actual numerical value of MEI of a pump size.

For both cases, the procedure of verification of the compliance or of the numerical value indicated by the manufacturer by institutions of market surveillance is described.

---

6. In this standard, relative flow rate is defined as the ratio of the actual flow rate to the flow rate at the operating point of best efficiency Q_{BEP}.

2.1.3 Standards Concerning Pump Units as Extended Products

The European Standard **EN 50598-1** [15] gives some general information on the method of determining energy efficiency indicators, as for example EEI, of motor-driven applications, by using the extended product approach and SAMs of a drive (motor or PDS) and a driven machine.

The special case of pump units as extended products (=combinations of drives and pumps) shall be covered by a new European standard. Work on this standard has already begun in the responsible Standardization Committee CEN/TC197/WG1. The provisional title of the standard is "Quantification of the energy efficiency of water pump units." It is intended to split the standard into three parts:

- Part 1: General description of the methodology
- Part 2: Single pump units
- Part 3: Multiple pump units (planned).

As preparation for the work of the Standardization Committee, draft proposals for Parts 1 and 2 have been elaborated and adopted in the EUROPUMP Joint Working Group on Energy-using Products (see Chapter 3).

In the draft proposal of Part 1, the EEI for pump units is defined and explained together with the definition of reference flow-time profiles and reference pressure control curves to be applied for various cases of typical pump operation (details are presented in Chapter 8). Also described in a general way are procedures to be applied

- in the frame of the so-called qualification by manufacturers or assemblers of pump units (i.e., determination of EEI in order to prove compliance with requirements by law and/or to indicate the actual value on the pump unit and in its documentation)
- in the frame of the so-called verification by institutions of market surveillance in order to verify the conformity or the indicated numerical value of pump units.

The draft proposal of Part 2 specifies—for the case of units consisting of one single pump with its drive (motor or PDS)—the reference power input $P_{1,\mathrm{ref}}$, the reference flow-time profiles, and the reference pressure control curve to be applied for the determination of EEI. Furthermore, two methods to determine EEI are described in detail:

- The experimental determination by tests
- The purely mathematical determination by application of SAMs of the individual components (pump and motor or PDS).

For the second method, Part 2 contains all the necessary information on the SAMs and their application (for the details, see Chapter 8).

The content of the planned Part 3 is still open and is being prepared by current work in the EUROPUMP Joint Working Group on Energy-using Products.

2.2   LEGISLATION IN EUROPE

Based on the EuP Directive 2005/32/EC [19], recast by 2009/125/EU, the European Commission (EC) defined a Lot 11 for implementing the directive in the field of electric motors and other equipment using electric motors (pumps, compressors, fans). Preparatory studies for electric motors [8], circulators for heating systems [7], and clean water pumps [6]—each of the products within a range of properties and fields of application—were procured by tender by the EC and finalized by the respective consultants in 2008.

2.2.1   EC Regulation for Electric Motors

Following the preparatory study for electric motors [8], the EC adopted in 2009 the EC Regulation [20], which implements the eco-design criteria for electric motors. This regulation imposes nominal minimum efficiency requirements for electric motors.

Included in the scope are squirrel-cage single-speed three-phase induction motors for 50 Hz or 50/60 Hz that

- have up to 6 poles
- have a rated voltage U_N of up to 1000 V
- have a rated output of $0.75 \leq P_N \leq 375$ kW
- are rated on the basis of continuous duty operation.

The regulation does not apply to various special motor types, for example,

- motors designed to operate wholly immersed in a liquid
- motors completely integrated into a product (e.g., gear, pump, fan, or compressor) of which the energy performance cannot be tested independently from the product.

According to the regulation,

- from January 1, 2015, motors with a rated output 7.5 kW−375 kW shall not be less efficient than defined by the efficiency level (=class) IE3 or meet the efficiency level (=class) IE2 and be equipped with a variable-speed drive
- from January 1, 2017, all motors with a rated output 0.75 kW−375 kW shall not be less efficient than defined by the efficiency level (=class) IE3 or meet the efficiency level (=class) IE2 and be equipped with a variable-speed drive.

The minimum efficiency values of electric motors for the respective efficiency levels (=classes) are defined in Annex I of Ref. [20], which is based on the values defined in Ref. [11]; see also Section 2.1.1.

It is important to note that the classification of electric motors in Ref. [20] and Ref. [11] takes only account of the motor efficiency at the nominal operating point, specified by the rated (=nominal) shaft power $P_{M,N}$, but not the motor efficiency at part load operation $P < P_{M,N}$ which is of at least comparable relevance for motors as components of pump units.

Concerning the experimental determination of the motor efficiency for the purposes of compliance and verification of compliance with the requirements of the regulation [20], in the Commission Communication 2012/C 402/07 [25] dated 20.12.2012 reference is given to the international standard [12], which is concerned with the measurement of motor efficiency.

In a Commission Working Document [21], the scope is intended to be enlarged to the range of rated output power $0.12 \leq P_{N,M} \leq 1000$ kW. Furthermore, the requirements regarding the efficiency classes of motors or alternatively their combination with a variable-speed drive shall be tightened step by step from the date of coming into force of the revised regulation until January 1, 2018. As a new item, the revised regulation shall contain the requirement that also variable-speed drives (=CDMs) must achieve a minimum efficiency according to a certain efficiency level (=class). Envisaged in the document is the class IE1 for all variable-speed drives from the date January 1, 2018. The nominal minimum efficiency requirements for variable-speed drives shall be given in an annex of the revised regulation.

2.2.2   EC Regulation for Circulators

Based on the preparatory study for circulators [7], the EC adopted in 2009 the EC Regulation [22] that implements the eco-design criteria for circulators in buildings. The regulation defines circulators that are concerned as glandless impeller pumps with a rated hydraulic output power of 1−2500 W, which are designed for use in heating systems or in secondary circuits of cooling distribution systems. As in the European Standard EN 16297 [16], a circulator is defined as being glandless if its motor shaft is directly coupled to the impeller and the motor is immersed in the pumped medium.

Two types of circulators are concerned:

- stand-alone circulators designed to operate independently from the product
- circulators integrated in products.

The regulation does not set efficiency requirements

- for drinking water circulators
- for a limited time to circulators integrated in products as replacement for identical circulators.

The energy efficiency requirements are stated in the following way:

- From January 1, 2013, glandless stand-alone circulators, with the exception of those specifically designed for primary circuits of thermal solar systems and of heat pumps, shall have an EEI of not more than 0.27.
- From August 1, 2015, glandless stand-alone circulators and glandless circulators integrated in products shall have an EEI of not more than 0.23.

The benchmark for the best available technology on the market for circulators is quoted to be EEI ≤ 0.20 at the time of the adoption of the regulation.

The determination of the EEI of circulators is described in Annex II of the regulation, which corresponds to the standard [16].

Requirements (based on a suggestion of EUROPUMP) concerning clarifications and amendments of the regulation [22] were stated in an EC Working Document [23] and have been implemented in the revised version of the Regulation (EU) No 622/2012 of July 11, 2012.

2.2.3   EC Regulation for Water Pumps

Based on the preparatory study for water pumps [6], the EC adopted in 2009 the EC Regulation [24] that implements the eco-design criteria for a range of pumps used in commercial buildings, drinking water, agriculture, and the food industry.

Concerned are rotodynamic water pumps for pumping clean water, including where integrated in other products.

The regulation does not apply to special types of rotodynamic pumps, for example,

- those designed for pumping clean water at extremely low or high temperatures (below −10°C or above 120°C) or for fire-fighting applications
- self-priming water pumps.

The pump types that are in the scope of the regulation [24] are

- glanded end-suction pumps
  - with own bearing (ESOB)
  - close coupled to the motor (ESCC)
  - close coupled to the motor with in-line flanges (ESCCi)
- glanded vertical multistage pumps (MS-V)
- submersible multistage pumps (MSS).

For each of these pump types, the scope of the regulation [24] is limited to certain ranges of design and nominal operating data and/or of pump sizes.

The ranges of design and/or nominal operating data are

- for the end-suction pumps:
  - design pressure up to 16 bar
  - specific speed n_s between 6 and 80 1/min
  - minimum rated flow rate 6 m³/h
  - maximum shaft power of 150 kW
  - maximum head 90 m at nominal speed 1450 1/min and maximum head 140 m at nominal speed 2900 1/min
- for the vertical multistage pumps
  - design pressure up to 25 bar
  - nominal speed 2900 1/min
  - maximum flow rate 100 m³/h
- for the submersible multistage water pumps
  - nominal outer diameter of 4″ or 6″ designed to be operated in a borehole
  - at a nominal speed 2900 1/min
  - at operating temperatures within a range of 0°C and 90°C.

The energy efficiency requirements are stated in the following way:

- From January 1, 2013, water pumps shall have a minimum efficiency corresponding to an MEI of 0.1
- From January 1, 2015, water pumps shall have a minimum efficiency corresponding to an MEI of 0.4.

The calculation of minimum efficiency is defined in Annex II, point 1(a) and in Annex II of the regulation [24] which corresponds to the European Standard prEN 16480 [18].

Furthermore, the regulation [24] establishes requirements regarding information to be given on the respective water pump by its manufacturer.

At the time the regulation entered into force, the corresponding European Standard prEN 16480 [18] had not been finally published. Therefore, in a Commission Communication [25], the needed information for the verification of compliance with the requirements of the regulation [24] was given. Concerning tests regarding the pump efficiency for the purpose of compliance with the requirements of the regulation [24], reference is made to the International Standard ISO 9906 [17]. It is requested that these measurements shall be done according to ISO 9906, Class 2b, with the exception that the total tolerance of pump efficiency for pumps having a mechanical power input of $P_{mech,N} < 10$ kW shall not be considered.

The Commission Communication [25] will be substituted after publication of the harmonized European Standard EN 16480 and publication of the title in the Official Journal of the European Union.

To extend the area of application of the EcoDesign Directive [19] in the fields of pumps and motors, the additional lots 28, 29 (focusing on pumps), and 30 (focusing on electric motors) were defined by the EC.

Lot 28 was originally defined to comprise "Pumps for private and public wastewater and for fluids with high solids content."

Lot 29 was originally defined to comprise "Pumps for private and public swimming pools, ponds, fountains, and aquariums (and clean water pumps larger than those regulated under ENER Lot 11)."

Lot 30 is intended to encompass products in motor systems outside the scope of the motor Regulation 640/2009 [20]. This includes, for example, other induction motors (single-phase and three-phase outside the $0.75-375$ kW power range), motors specifically designed for being fed by a frequency converter, and permanent magnet motors. Also included are motor drives and controllers as electro-mechanical starters, soft starters, and VSDs.

For these lots, preparatory studies were carried out, and the final reports [26−28] were published in 2014. A legislative implementation of lots 28, 29, and 30 is presently still in discussion among the EC and concerned parties.

2.2.4 Concluding Remarks

All three existing Commission Regulations [20,22,24] that are of concern for the energy efficiency of pump units and already entered into force define requirements on single products (motors, circulators, pumps). They can be assigned to the *Product Approach* on energy efficiency. The typical operation mode of these products (in the case of motors as components of pump units), characterized by running (far) apart from the nominal conditions during a considerable part of operating hours, is

- principally not reflected by the requirements on the efficiency of electric motors
- reflected by the requirements on the efficiency (via the MEI) of water pumps only for operation in a relatively narrow range of flow rates near BEP
- actually reflected by the requirements on the EEI of circulators.

The *Extended Product Approach* for pump units, which principally takes account of their typical operation modes, is already addressed and foreseen in article 7 (Revision) of the regulation [24] for water pumps and is coming to be addressed by the studies [26] and [27] according to the new lots 28 and 29.

Chapter 3

Overview on Europump Approaches on Assessment of Energy Efficiency

Europump, the European Association of Pump Manufacturers, and its members have been deeply involved for many years in various activities that aim to create methods, tools, and classes or indicators for assessing the energy efficiency of pumps and pump units.

As an early reaction to the European goal of saving energy by improving the energy efficiency of energy using products, European manufacturers of circulators agreed on a voluntary commitment [29]. It became active in March 2005. The main objective of this commitment was to determine the principles of a classification scheme related to energy labeling in order to achieve major savings in energy consumed by circulators. The commitment encompassed stand-alone circulators with a nominal electric power ≤ 2500 W. To supervise the implementation into the products a steering committee composed of signatories of the commitment and Europump representatives was established with the responsibility of handling all administrative matters and of reporting annually to the European Commission about proceedings.

In the technical part of the commitment, efficiency classes are defined as alphabetical characters from G to A (with G being the worst one) corresponding to ranges of an Energy Efficiency Index (EEI). For the latter, its definition including a reference load ($=$flow-time) profile, a reference pressure control curve, and a reference power as well as the method for the determination of the EEI (as defined in Ref. [29]) by tests and evaluations was described in the document. Even though this voluntary commitment was later substituted by the EC Regulation for circulators [22] (see Section 2.2), it formed the direct basis for the introduction of the EEI—in a very similar manner with only slight modifications—into the legislation for circulators. An important contribution of Europump to the content of the regulation [22] as well as to the European standard [16] for circulators consisted in the collection and evaluation of data regarding the energy efficiency of state-of-the-

Assessing the Energy Efficiency of Pumps and Pump Units.
DOI: http://dx.doi.org/10.1016/B978-0-08-100597-2.00003-3
© 2015 Elsevier Ltd. All rights reserved.
41

art circulators on the market that served to derive equations for the reference power P_{ref} as part of the definition of EEI (see Chapter 8).

Lot 11—as defined by the EC—comprises "energy using products." Circulators as one category concerned by lot 11 consist of a pump part, a motor, and—as an advanced version—of a VSD and an electronic controller. But these parts are mechanically and functionally integrated and are produced and sold by one manufacturer as one single product. On the other hand, pumps of the types covered by the preparatory study [6] and the EC Regulation [24] in the frame of lot 11 and of the European standard [18]—even though they can only be operated in combination with a drive (motor or PDS)

- are addressed in these documents as separate products
- shall be assessed in respect to their efficiency independently from their drives and modes of operation
- and are objects of the Product Approach (PA).

Therefore, Europump started early with activities to support work on the preparatory study [6] and on the regulation [24]. For this purpose, Europump established a "Joint Working Group on Energy using Products (JWG on EuP)" with some subgroups of experts for special topics.

In this group, the concept of the Minimum Efficiency Index (MEI) was developed (described in Chapter 7). A basic principle of this concept is to include into the assessment of the energy efficiency of pumps operated at fixed speed—besides their efficiency at nominal operating condition (BEP)—also the pump efficiency in a certain range around BEP.

To create a scientific and technical basis for a method of assessing the energy efficiency of *pumps* by their MEI, a project at the Technische Universitaet (=University of Technology) Darmstadt (TUD) was initiated and supported by Europump; see the final report [30]. This project aimed at

- the collection and evaluation of data concerning the efficiencies of state-of-the-art pumps on the European market in relation to pump attributes (as type, speed, and others)
- the elaboration of appropriate correlations that can be used for assessing and legislative regulation of the pump efficiencies.

Based on the results of this project and extensive discussions in the JWG, a proposal for the European standard on MEI [18] was worked out and transmitted as input to the responsible Standardization Committee.

After finishing the work done by Europump on the Product Approach (PA) for clean water pumps and on the concept of MEI, the activities of Europump, organized by the JWG, were focusing on the Extended Product Approach (EPA) for pump units. Because of the inclusion of electric components (motor or PDS) in the treatment of pump units as extended products, from the beginning of the work on this item experts from the side of motor, converter, and PDS manufacturers became regular members of the expert subgroups.

Vice versa, Europump has been involved in the work done by the responsible committee of CENELEC on the elaboration of the European standard [15]. This involvement has been by direct participation of representatives of the JWG, by a permanent exchange of information between both groups, and by delivering contributions in respect to pump units serving as an application example.

Based on the well-established use of the EEI as an energy efficiency indicator for circulators, it was an early decision taken by the JWG of Europump to follow the EEI concept also in the frame of the EPA for pump units. This decision was confirmed by the parallel work within CENELEC on the standard [15], which also implied the EEI as a general efficiency indicator for extended products.

To create also a solid scientific and technical basis for a method of assessing the energy efficiency of *pump units*, again a project was initiated and supported by Europump and carried out at the TUD. This extensive project consisted of an experimental and a theoretical part; see the final reports [31] and [32]. It aimed at

- proving an experimental method for the determination of the EEI of single pump units with AC induction motors, investigating the influence of various attributes of pump units, and to provide test data for the validation of a method for the determination of EEI by applying semi-analytical models (SAMs)
- the development, the validation, and the application of SAMs enabling the determination of the EEI of single pump units by calculations.

Based on the results of this project and on extensive discussions in the JWG draft proposals for Parts 1 and 2 of a European standard on EEI for pump units (already mentioned in Section 2.1) have been elaborated and adopted in the JWG.

The details of the methodology of EEI applied to the assessment of the energy efficiency of pump units are explained in Section 8.3 of this book.

It should be mentioned that Europump—in addition to the work of the JWG described before—provides information in the form of Europump guides for companies and institutions that are affected by existing and expected legislative measures in respect to requirements on energy efficiency of their products. The guideline [33] concerns aspects in relation to the practical application of the EC Regulation for clean water pumps [24]. A guide that concerns aspects of the EPA is being prepared and exists already in a draft version [34].

Chapter 4

Physical and Technical Background of the Efficiency of Pumps

4.1 PUMP EFFICIENCY AT THE BEST EFFICIENCY POINT (BEP)

The definition of the efficiency of a pump was already given in Section 1.2. By combining Eqs (1.14) and (1.15) of that chapter, the pump efficiency is also determined by the equation

$$\eta_{\text{pump}} = \frac{\rho \cdot g \cdot Q \cdot H}{P_{\text{mech}}} \tag{4.1}$$

where Q, H, and P_{mech} are the flow rate, the head, and the mechanical power input of the pump, respectively. In practice, the efficiency is determined experimentally by measuring—with the aid of suitable measuring equipment—the flow rate, the head, and the mechanical power input and calculating the efficiency by Eq. (4.1). The gravitational constant g is 9.81 m/s^2. In the case of tests with clean cold water the liquid density ρ can be set with sufficient accuracy to 1000 kg/m^3. Typically, the efficiency is determined at a number of test points (i.e., values of the flow rate) at constant rotational speed n,[1] normally at the nominal rotational speed $n_{\text{pump,N}}$ of the pump, which is for example 2900 1/min for pumps designated to be driven by a two-pole 50 Hz AC induction motor.

Inevitable small fluctuations of the measured quantities during the tests due to the characteristics of the measuring system or to random variations of the measured quantities cause slightly different values of the instrument readings of the same quantity at the same test point. This is called random

---

1. The actual speed is also measured for each test point. Inevitable small deviations from the intentionally constant speed during the test are acceptable and the measured values of Q, H, and P_{mech} can be corrected by the means of the hydrodynamic laws of similarity.

Assessing the Energy Efficiency of Pumps and Pump Units.
DOI: http://dx.doi.org/10.1016/B978-0-08-100597-2.00004-5
© 2015 Elsevier Ltd. All rights reserved.

uncertainty of the measured quantities.[2] By the "law of error propagation" the random uncertainties of the individual measured quantities in Eq. (4.1) are augmented to the resulting random uncertainty of the pump efficiency. Therefore, the values of the efficiencies determined experimentally at the test points and plotted versus the corresponding measured values of the flow rate show typically a certain scatter. This is illustrated exemplarily in Figure 4.1.

However, test results show generally the pump efficiency to increase with increasing flow rate (beginning from zero at zero flow rate), to reach a maximum and then to drop down with further increasing flow rate. The operating point according to the maximum efficiency is called "best efficiency point" (BEP), the corresponding values of flow rate and pump efficiency are named Q_{BEP} and η_{BEP}, respectively. The value Q_{BEP} is usually also declared as the *nominal* value of the flow rate of the respective pump. The same holds true for the values H_{BEP} and $P_{mech,BEP} = P_{BEP}$ that correspond to Q_{BEP}. Usually the operating range $Q < Q_{BEP}$ is called "part load range," while the range $Q > Q_{BEP}$ is called "overload range."

Because of the scatter of the points in the diagram, the actual values of Q_{BEP} and η_{BEP} are difficult to identify exactly from only the measured and plotted points. Therefore, the test points are usually approximated by a mathematical best-fit function, mostly in the form of a polynomial as shown in

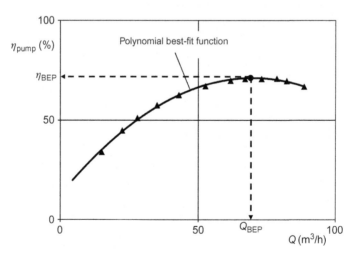

FIGURE 4.1   Exemplary determination of Q_{BEP} and η_{BEP} from a test.

---

2. Requirements regarding maximum allowable values of fluctuations of measured quantities are defined by the relevant test Standard ISO 9906 [17] for the respective grade of accuracies and tolerances.

Figure 4.1. From this polynomial, the values of Q_{BEP} and η_{BEP} can be calculated exactly.

Relating to the internal power losses in a pump, the pump efficiency can also be expressed by the following equation:

$$\eta_{pump} = \frac{P_{mech} - \sum P_{loss,pump}}{P_{mech}} = 1 - \frac{\sum P_{loss,pump}}{P_{mech}} = 1 - \frac{\sum P_{loss,pump}/P_{BEP}}{P_{mech}/P_{BEP}}$$

$$(4.2)$$

In this equation, P_{loss}/P_{BEP} are related power losses (=related to the nominal pump power input) that occur in the pump and P_{mech}/P_{BEP} is the relative pump power input. Especially for the point of BEP, the relative pump power input is equal to 1 and the efficiency is maximum ($\eta = \eta_{BEP}$).

The sum of pump power losses comprises various individual losses arising in the pump. These are

● mechanical friction losses

$$P_{loss,mech} = \omega \cdot \sum T_{mech,friction} \qquad (4.3)$$

● volumetric losses

$$P_{loss,volumetric} = \rho \cdot g \cdot \sum Q_{loss} \cdot H \qquad (4.4)$$

● through-flow losses

$$P_{loss,through-flow} = \rho \cdot g \cdot Q \cdot \sum H_{loss,pump} \qquad (4.5)$$

● disc friction losses

$$P_{disc\ friction} = \omega \cdot \sum T_{disc\ friction} \qquad (4.6)$$

Concerning the relative magnitude of the individual losses, the through-flow losses and the disc friction losses are the prevailing ones—except for pumps of very small size—for pump types, sizes, and shapes covered by the standardization and legislation addressed in Chapter 2. Under the same constraints in respect to considered pump types, sizes and shapes, volumetric losses, and mechanical friction losses are of minor magnitude compared to the other losses.

The *mechanical friction losses* result from friction torques $T_{mech,friction}$ caused by friction in the shaft sealing(s) and in the shaft bearings. Circulators have neither a shaft sealing nor their own pump bearings and, therefore, are not affected by mechanical losses. The pump types ESCC and ESCCi (see Section 2.2) have no own bearings and, therefore, are not affected by bearing friction losses. The magnitude of the *related mechanical friction losses* is directly determined by the type and the design details of the shaft sealing (stuffing box, mechanical seal) and of the shaft bearings (roller bearings, sliding bearings). The contribution of the mechanical friction losses to the total power losses of pumps increases with decreasing pump size and becomes of considerable importance for pumps (with shaft seals and own bearings) of very small size.

The *volumetric losses* result from internal leakage in the form of (small) flow rates $\sum Q_{loss}$ recirculating through internal sealing gaps at the impeller suction eye and—in the case of multistage pumps—between shaft and stationary parts as well as through gaps and/or holes that serve for axial thrust balancing.

Regarding the absolute magnitude of the *internal leakage* $\sum Q_{loss}$, it depends on

- the pressure rise generated by the impeller(s)
- the throttling resistance of the internal sealing gaps, particularly on their gap width.

In a first approximation (neglecting the radial pressure variation in the spaces between impeller shrouds and casing walls caused by fluid rotation), the internal leakage can be assumed to be proportional to the square root of the impeller pressure rise, which for its part is closely connected to the pump head (in the case of multistage pumps divided by the number of stages). For pumps of different size, the ratio of gap width(s) to impeller diameter is important for the magnitude of the internal leakage and, thereby, of the volumetric power losses. The absolute gap width cannot be decreased proportionally to the pump and impeller size for reasons of manufacturing and avoidance of rubbing when running. Therefore, the related volumetric losses increase typically with decreasing pump size and attain enhanced importance for pumps of very small size.

The totality of *through-flow head losses* $\sum H_{loss,pump}$ consists of

- *wall friction losses*
- *shock (or incidence) losses*
- *mixing (or wake) losses*
- *part-load recirculation losses.*

Shock losses result from deviation of the flow direction at the inlet of the impeller and of the volute or guide vanes from the optimal (="shockless") direction.

Mixing losses are caused by energy dissipation in flow zones of nonuniform velocity distribution downstream of the impeller outlet.

Part-load recirculation losses are caused by energy dissipation in flow recirculation zones at the inlet and/or outlet of the impeller. These recirculations occur only below critical values of the flow rate.

At the BEP the part-load recirculation losses are zero and the shock and mixing losses are very small compared to the wall friction losses for pumps that have a well-designed geometry—under hydrodynamic aspects—of the impeller blading and of the stationary parts along the flow path.

The *wall friction part* of the through-flow losses results from viscous friction of the trough-flow along the wetted surfaces of impeller and stationary pump parts (i.e., volute in the case of end-suction pumps or guide vanes and return channels in the case of multistage pumps, also casing inlet and outlet). The through-flow losses by wall friction can be considered in analogy to the friction loss in a straight pipe. For the latter, the head loss is described by the following equation:

$$H_{loss,pipe} = \lambda \cdot \frac{L_{pipe}}{D_{pipe}} \cdot \frac{c_{pipe}^2}{2g} \qquad (4.7)$$

In this "pipe friction equation," λ is the dimensionless pipe friction coefficient, L_{pipe} is the length, and D_{pipe} is the inner diameter of the pipe, respectively, and c_{pipe} is the mean flow velocity along the pipe axis ($=$ transport velocity). Equation (4.7) can be converted to the form

$$H_{loss,pipe} = \lambda \cdot \left(\frac{4}{\pi}\right)^2 \cdot \frac{L_{pipe}}{D_{pipe}^5} \cdot \frac{Q^2}{2g} = \lambda \cdot SF_{pipe} \cdot \frac{Q^2}{2g}. \qquad (4.8)$$

The shape factor SF_{pipe} represents the pipe geometry (described by L_{pipe} and D_{pipe}).

According to the well-known COLEBROOK diagram (and the corresponding mathematical approximation equations), the friction coefficient λ for pipes with technically rough inner surface depends

- on the (dimensionless) pipe Reynolds number

$$Re_{pipe} = \frac{c_{pipe} \cdot D_{pipe}}{\nu} = \frac{4}{\pi} \cdot \frac{Q}{D_{pipe} \cdot \nu} \qquad (4.9)$$

with the kinematic viscosity ν of the fluid
- and on the relative roughness k/D_{pipe} of the wetted surface with k as a characteristic size of the surface roughness.

The correlation is illustrated exemplarily for two values of the relative roughness in Figure 4.2.

Transferred to the head loss by wall friction in pumps, it can be approximatively described by

$$\sum H_{loss,friction} = \lambda_{pump} \cdot SF_{pump,Q} \cdot \frac{Q^2}{2g}. \qquad (4.10)$$

The shape factor $SF_{pump,Q}$ is then determined by the shape of the wetted surfaces of impeller blading and of all stationary parts along the flow path from pump inlet to outlet.

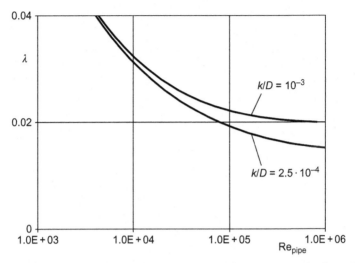

FIGURE 4.2 Pipe friction coefficient λ in dependence of the Reynolds number Re_{pipe} for two exemplary values of the relative surface roughness k/D_{pipe}.

The through-flow friction coefficient λ_{pump} for pumps is an average value along the flow path from pump inlet to outlet and depends on a (characteristic) pump Reynolds number and on the average relative surface roughness $k_{average}/D_{impeller}$ along the flow path from pump inlet to outlet.

The pump Reynolds number is usually defined by

$$Re_u = \frac{u_{impeller} \cdot D_{impeller}}{\nu} \tag{4.11}$$

$D_{impeller}$ is the outer diameter of the impeller and $u_{impeller}$ is the circumferential velocity of the impeller at its outer diameter.

In respect to the through-flow and corresponding friction losses in the pump, a Reynolds number defined by

$$Re_Q = \frac{Q_{BEP}}{D_{impeller} \cdot \nu} \tag{4.12}$$

is more adequate, and the (average) friction coefficient is then primarily dependent on Re_Q and on the relative surface roughness.

The *disc friction losses* result from friction torques $T_{disc\ friction}$ caused by viscous wall friction on the wetted outer surfaces of all rotating parts, mainly of the front and back shrouds of the impeller(s). They can be considered in analogy to the viscous friction exerted on a plane disc rotating in a liquid and to the resulting frictional torque. This frictional torque is usually expressed by an equation of the type

$$T = c_T \cdot SF_{disc} \cdot \left(\frac{D_{disc}}{2}\right)^3 \cdot \frac{\rho}{2} \cdot u_{disc}^2 \tag{4.13}$$

In this "disc friction equation," c_T is a dimensionless friction torque coefficient, D_{disc} is the outer diameter of the disc, and u_{disc} is the circumferential disc velocity at the outer disc diameter. The shape factor SF_{disc} depends on the geometry of the disc (thickness to diameter ratio) and of its surrounding (e.g., the ratio of the gap width between disc and a cylindrical casing to the disc diameter).

The disc friction coefficient depends

- on the (dimensionless) disc Reynolds number

$$\text{Re}_{disc} = \frac{u_{disc} \cdot D_{disc}}{\nu} \tag{4.14}$$

- and on the relative roughness(es) k/D_{disc} of the wetted surfaces of the disc and (in the case of an enclosed disc) of the casing.

The dependency $c_T = f(\text{Re}_{disc}, k/D_{disc})$ for simple geometrical cases—for instance, for a plane disc without casing or for a plane disc in a cylindrical casing—looks qualitatively very similar to that for the pipe friction coefficient shown in Figure 4.2.

Transferred to the disc friction in pumps, it can be approximatively described by

$$\sum T_{disc\ friction} = c_{T,pump} \cdot SF_{pump,u} \cdot \left(\frac{D_{impeller}}{2}\right)^3 \cdot \frac{\rho}{2} \cdot u_{impeller}^2 \tag{4.15}$$

The shape factor $SF_{pump,u}$ is then determined by the shape of the wetted surfaces of the liquid-filled clearances between impeller(s) and casing.

The friction torque coefficient $c_{T,pump}$ is a value that includes contributions of all these wetted surfaces and depends on a (characteristic) pump Reynolds number and on the average relative roughness $k_{average}/D_{impeller}$ of the contributing surfaces. For the disc friction losses, the pump Reynolds number as defined by Eq. (4.11) is an adequate one.

From these analogies, direct conclusions can be drawn regarding the influence of pump size on the wall friction part of the related through-flow losses and on the related disc friction losses and, thereby, according to Eq. (4.2) on η_{BEP}. For pumps that are geometrically similar to each other, it follows from the laws of hydrodynamic similarity for corresponding relative flow rates Q/Q_{BEP}—and especially for $Q/Q_{BEP} = 1$

$$Q \sim n \cdot D^3 \tag{4.16}$$

In Eq. (4.16), D represents any characteristic length, normally the outer impeller diameter $D_{impeller}$ taken as a representative diameter. For the same rotational speed n of geometrically similar pumps, it follows from Eq. (4.16) that the size of the pumps is proportional to the cubic root of Q_{BEP}.

$$D_{impeller} \sim \sqrt[3]{Q_{BEP}} \tag{4.17}$$

Also, in the case of the same rotational speed n there is a simple proportionality

$$u_{impeller} \sim D_{impeller} \qquad (4.18)$$

Combining Eqs (4.17) and (4.18) with the definitions of the characteristic pump Reynolds numbers according to Eqs (4.11) and (4.12), respectively, it becomes evident that—for geometrically similar pumps operated at the same rotational speed—both Reynolds numbers are directly correlated with each other and with Q_{BEP} by the proportionalities

$$\text{Re}_u \sim \text{Re}_Q \sim Q^{2/3} \qquad (4.19)$$

For geometrically similar pumps operated at the same rotational speed, the dependence of the wall friction part of the related through-flow losses and of the related disc friction losses on the associated Reynolds numbers is, therefore, also reflected by their dependence on the nominal flow rate Q_{BEP}. The dependence on Q_{BEP} will be characterized by a decrease with increasing Q_{BEP} approaching asymptotically minimum values that, for their part, depend on the shape (factors) of the geometrical design and on the relative surface roughness.

Including the contribution of the related volumetric losses and of the related mechanical friction losses and their dependence on the pump size (explained above), the fundamental influence of Q_{BEP} on the related power losses for geometrically similar pumps operated at the same rotational speed is determined by physical and technological reasons. Of course, actual values of η_{BEP} of individual pumps are additionally dependent on their design and manufacturing quality in respect to energy conversion ($=$"efficiency related quality"), which is superposed to the fundamental influences on η_{BEP}. The latter appear in pure form only for pumps of comparable efficiency related quality. Collecting and evaluating data of pumps on the market can serve to prove the fundamental influences on η_{BEP} as well as to quantify the range of scatter of actual values of η_{BEP} resulting from differences in efficiency related quality; see for example Refs [30,35].

The evaluation is normally based on ordering the data according to the specific speed n_s of the pumps. The specific speed is defined by

$$n_s = n_N \cdot \frac{Q_{BEP}^{0.5}}{(H_{BEP}/i_{st})^{0.75}} \qquad (4.20)$$

In this (dimensional) definition equation, the unit of the specific speed n_s and of the nominal speed n_N is 1/min, the unit of the flow rate Q_{BEP} is m³/h and the unit of the pump head H_{BEP} is m. The integer number i_{st} is equal to 1 in the case of single-stage pumps and equal to the number of stages in the case of multistage pumps.

For geometrically similar pumps, it follows from the laws of hydrodynamic similarity for corresponding relative flow rates Q/Q_{BEP}—and especially for $Q/Q_{BEP} = 1$

$$H \sim n^2 \cdot D^2 \tag{4.21}$$

Therefore, geometrically similar pumps of different size (defined by the magnitude of D, normally of the outer impeller diameter $D_{impeller}$) and corresponding different values of Q_{BEP} at the same rotational speed n have the same specific speed n_s. Vice versa, but in a somewhat weaker strictness, pumps of the same type that have the same specific speed n_s are reasonably similar to each other in respect to their geometry, at least in their main dimensions of impeller and stationary parts along the flow path.

From this reason, pumps of nearly the same specific speed and of comparable efficiency related quality will reflect the fundamental influence of pump size (represented by Q_{BEP}) on the related losses at BEP. The latter can be derived from the actual values of η_{BEP} by the inverted form of Eq. (4.2) applied to BEP.

$$\frac{\sum P_{loss,BEP}}{P_N} = 1 - \eta_{BEP} \tag{4.22}$$

From the data collection and analysis reported in Ref. [30], the sum of related losses at BEP in dependence of Q_{BEP} can be evaluated and is presented here exemplarily for pumps

- of the end suction type with own bearings (ESOB) and a nominal speed of 2900 1/min
- of a medium quality of efficiency

for two values of the specific speed n_s (within the range, which is of interest here).

The result is shown in the form of best-fit curves in Figure 4.3 and confirms the fundamental dependence of the total related losses of pumps on Q_{BEP} as reasoned above.

This fundamental dependence of the total related losses on the flow rate Q_{BEP} determines directly the dependence of the efficiency η_{BEP} of pumps of a certain type, of a certain specific speed n_s, and with the same nominal speed n_N on Q_{BEP}. This is illustrated exemplarily in Figure 4.4 for the same pump type, nominal speed, and efficiency related quality as for the total related losses shown in Figure 4.3.

As it is already apparent from Figures 4.3 and 4.4, there is also an influence of the specific speed on the total related losses and on η_{BEP} for any value of Q_{BEP} at a certain nominal speed. The dependence on the specific speed results from the physical nature of the various related individual losses and from their relation to the geometry of the pumps. The latter is, for its part, related to the specific speed. To illustrate this relation, an example is

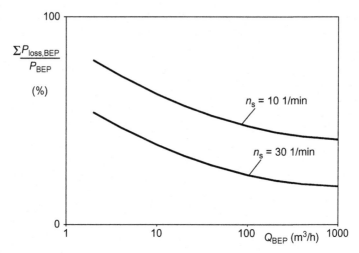

FIGURE 4.3 Example for the dependence of the total related losses $\sum P_{loss,BEP}$ on the flow rate Q_{BEP} at constant specific speed n_s.

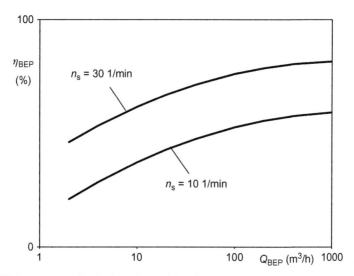

FIGURE 4.4 Example for the dependence of the efficiency η_{BEP} on the flow rate Q_{BEP} at constant specific speed n_s.

shown in Figure 4.5. This example, confirmed by results of a project at the TU Darmstadt reported in Ref. [39], applies to end-suction pumps with their own bearings, all having the same value of Q_{BEP} and the same nominal speed $n_{pump,N}$. It should be interpreted in a rather qualitative sense.

Obviously, the sum of the related losses is lowest for values of the specific speed in the range of 40–50 1/min. This characteristic of the total

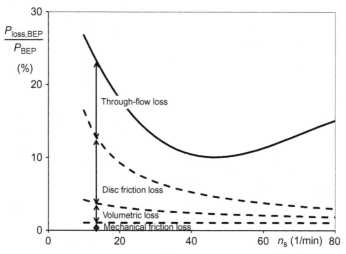

FIGURE 4.5 Fundamental dependence of the related losses $\sum P_{\text{loss,BEP}}$ on the specific speed n_s (example).

related losses results from the different dependencies of the various related losses on n_s, also plotted as dashed lines in Figure 4.5. The related mechanical friction losses (at least in this example) are very low and independent from n_s. The related volumetric losses as well as the related disc friction losses show a distinct increase with decreasing n_s. This is caused by the increase of pump head H_{BEP} and of the corresponding impeller diameter with decreasing n_s for constant flow rate Q_{BEP} and for constant nominal speed $n_{\text{pump,N}}$; see Eq. (4.20). The pump head is responsible for the magnitude of the internal leakage flow. The impeller diameter affects extensively the friction torque (according to Eq. (4.15) by the fifth power) for comparable values of the friction coefficient $c_{\text{T,pump}}$ and of the shape factor $\text{SF}_{\text{pump,u}}$. The related through-flow losses are responsible for the increase of the total related losses with increasing n_s in the range above the minimum. In spite of the fact that the *absolute* magnitude of the internal head losses tend to decrease with increasing n_s, the *related* through-flow losses decrease because of the extensive decrease of the nominal power P_{BEP} (which is inversely proportional to H_{BEP}) in the denominator of the *related* through-flow losses.

This fundamental dependence of the total related losses on the specific speed is also confirmed by the data collection reported in Ref. [30]. For illustration, the total related losses are exemplarily evaluated by Eq. (4.22) for the same category of pumps as in Figure 4.3 and for two values of the flow rate Q_{BEP} (within the range which is of interest here). The result is presented in Figure 4.6 in the form of best-fit curves.

The dependence of the efficiency η_{BEP} on the specific speed is shown for the same example in Figure 4.7.

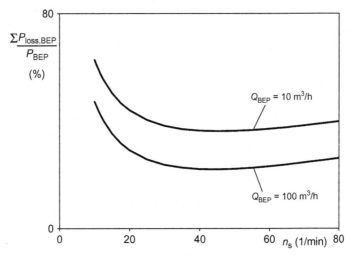

FIGURE 4.6   Example for the dependence of the total related losses $\sum P_{\text{loss,BEP}}$ on the specific speed n_s at constant flow rate Q_{BEP}.

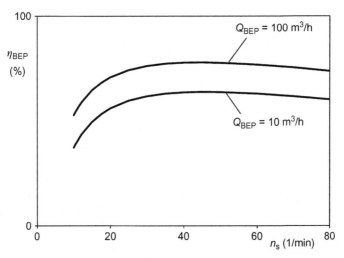

FIGURE 4.7   Example for the dependence of the efficiency η_{BEP} on the specific speed n_s at constant flow rate Q_{BEP}.

Besides its fundamental dependence on pump type, nominal speed $n_{\text{pump,N}}$, flow rate Q_{BEP}, and specific speed n_s, the actual efficiency η_{BEP} of any pump (or any size of pumps manufactured in series) depends on the *efficiency related quality* (see for example Ref. [38]). The latter results from

- the design of the inner pump geometry, which is relevant for energy conversion and losses

- the roughness of the wetted surfaces, which is relevant for friction losses
- possible deviations of the actual inner geometry from that defined by the design.

The shape factors (see Eqs (4.10) and (4.15)) that are relevant for the friction losses, but also for the magnitude of the shock and mixing losses, are determined by the design of the inner geometry. The "design quality" is based on the knowledge and experience of the pump manufacturer.

The roughness of relevant surfaces (impeller blade channels, volute or guide vanes, outer surfaces of impeller shrouds, and opposite inner casing surfaces) is determined by

- the material the respective pump parts are made of (e.g., cast iron, alloys, steel sheets, plastics)
- the processes of manufacturing the pump parts (casting, CNC milling, welding of preformed steel sheets, injection molding of plastics) and/or finishing the surfaces (milling, turning, polishing, coating).

In an earlier study at the TU Darmstadt [39], values of η_{BEP} were derived that are *only theoretically attainable* under idealizing assumptions (e.g., no restrictions for the design by other requirements on the pump or by material and manufacturing aspects, all wetted surfaces hydraulically smooth, minimum realizable gap clearances) for various pump types. However, in reality restrictions exist for the hydrodynamic and mechanical design of pumps because of

- competing requirements that are in conflict with maximizing the pump efficiency, as are requirements on cavitation performance, on mechanical stiffness, on low mechanical vibrations and/or on low noise, preferred construction materials, and manufacturing methods
- economic reasons in respect to material and manufacturing costs in relation to prices that conform to the market.

Therefore, the *practically and commercially attainable* values of η_{BEP} can never reach the theoretically maximum attainable ones.

Some information on practically realizable measures to improve η_{BEP} and on the resulting quantitative improvements can be found in Refs [37,40].

Generally, improving the efficiency related quality of pumps requires effort and costs for redesigning and superior materials and/or manufacturing processes.

4.2 PUMP EFFICIENCY APART FROM THE BEP AND AT VARIABLE SPEED

Even for pump applications that require only variations of the flow rate in a relatively small range below or around BEP and that should be operated preferably at fixed pump speed, the variation of the pump efficiency within the range of the relative flow rate Q/Q_{BEP} from about 75% to 110% is of interest.

For pump units that are operated at variable speed, the value of Q_{BEP} varies according to the similarity law for the flow rate (Eq. (4.16)) proportionally to the pump speed. The relative flow rate Q/Q_{BEP} is then the ratio of the actual value of Q at any pump speed to the value of Q_{BEP} at the same pump speed. The pump speed and the corresponding value of Q_{BEP} vary with the demanded flow rate Q. Depending on the pump $Q-H$ characteristic and for typical flow-time profiles and pressure control curves (see Chapter 8), the pump speed and Q_{BEP} vary from about 70% to 100% of their maximum values, but the corresponding values of the related flow rate Q/Q_{BEP} vary in a quite wider range. Therefore, the variation of the pump efficiency in this whole range of the relative flow rate is relevant for pump units operated at variable speed. However, it must be emphasized that the effect of speed variation on the mechanical power input outbalances typically the corresponding variation of the pump efficiency by far (see Section 5.2 and Chapter 8).

Generally, the variation of the relative pump efficiency η/η_{BEP} with the relative flow rate Q/Q_{BEP} is determined by the dependence of the individual and total related losses on the relative flow rate Q/Q_{BEP}. The related losses caused by *mechanical friction* and *disc friction* are not or only weakly dependent on Q/Q_{BEP}. For the two other related losses only some general or simplified correlations can be used for an explanation. The variation of the related *volumetric losses* is related to the variation of the relative pump head H/H_{BEP} with the relative flow rate Q/Q_{BEP} (the so-called slope or steepness of the $Q-H$ characteristic) which for its part is in a fundamental manner related to the specific speed. Concerning the related *through-flow losses*, it can be assumed in a considerably simplifying manner that

- the *head losses* caused by *wall friction* vary with the square of the flow rate (see Eq. (4.10)) and the resulting power losses vary with Q^3 (according to Eq. (4.3))
- the *head losses* caused by *"shock"* vary (very approximately) with the square of the difference $Q - Q_{BEP}$ and the resulting power losses vary (according to Eq. (4.3)) with $Q \cdot (Q - Q_{BEP})^2$.

On the other hand, the "coefficients" of this approximate relation as well as the contributions of the head losses by mixing and part load recirculations depend on the geometry of the wetted surfaces along the flow path from pump inlet to outlet that are related—in a more fundamental sense—to the specific speed and result—last but not least—from the specific features of the design executed by the responsible engineers. This design is typically based on a general "design philosophy" of the pump manufacturer. For illustration, the ratio of the total losses to their respective values at BEP is evaluated for three pumps out of the pumps tested in an experimental project at the TU Darmstadt [31] and plotted versus the relative flow rate Q/Q_{BEP} in Figure 4.8.

For the same three exemplary pumps, Figure 4.9 shows the relative efficiency η/η_{BEP} in dependence of the relative flow rate Q/Q_{BEP}.

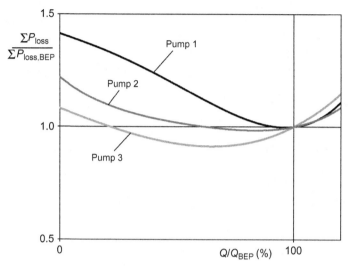

FIGURE 4.8 Dependence of the relative total losses on the relative flow rate Q/Q_{BEP} for three exemplary pumps.

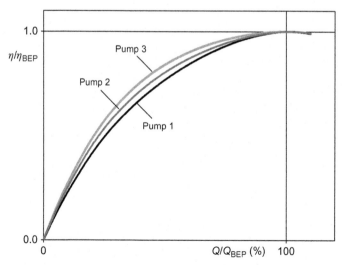

FIGURE 4.9 Dependence of the relative efficiency η/η_{BEP} on the relative flow rate Q/Q_{BEP} for three exemplary pumps.

The characteristics shown in Figure 4.9 are representative for the maximum variation of the shape of the $Q-\eta$ curves within the range of pump types and pump nominal data that are of interest here. The differences are most pronounced around $Q/Q_{BEP} = 50\%$ and of minor magnitude in the range of $75\% \leq Q/Q_{BEP} \leq 110\%$, which is relevant for fixed-speed operation near BEP (and for the concept of MEI described in Chapter 7).

In the case of *variable speed operation*, the pump Reynolds number as defined by Eqs (4.11) or (4.12) varies proportionally to the pump speed n. Because of the dependence of part of the related losses on the Reynolds number, in principle an influence of the pump speed on the pump efficiency should be available. Another effect may arise from the variation of the (absolute magnitude of the) mechanical friction losses that will normally show a different dependence on the pump speed than the pump power input P_{BEP} that it is related to.

Both the *related* losses caused by viscous wall friction and by mechanical friction tend to increase with decreasing pump speed. As a consequence, the pump efficiency tends to decrease with decreasing pump speed.

However, the actual effect for pumps operated at variable speed is relatively marginal or even negligible from two reasons:

1. As explained before, the range of variation of the pump speed is relatively small (in the order of only about 30% of the nominal speed for typical pressure control curves).
2. The Reynolds numbers of pumps at 100% pump speed are mostly in a range where the variation of the friction coefficients vary only slightly with the Reynolds number (see the fundamental dependence shown in Figure 4.2).

An exception to the second reason exists in the case of pumps that are relatively small in respect to their size and nominal flow rate. For such pumps, the Reynolds numbers may be small enough to be in the range where its influence on the friction coefficients is more pronounced.

Generally, the effect of Reynolds number variation (by size, speed, and/ or viscosity) on the efficiency of pumps (and other turbomachines) is usually called "scale-up effect." For the quantitative determination, often an equation (called ACKERET scaling-up formula) of the type

$$\frac{1 - \eta_2}{1 - \eta_1} = V \cdot \left(\frac{Re_2}{Re_1}\right)^{-\alpha} + (1 - V) \tag{4.23}$$

with $0 \leq V \leq 1$ is used. This equation expresses the fact that a part "V" of the related losses $(1 - \eta)$ depends on the Reynolds number by a power law with negative exponent while the part $1 - V$ is independent from the Reynolds number.

The evaluation of test results in the frame of the project at the TU Darmstadt reported in Ref. [31] showed that for the range of pumps covered by the types and nominal data of the test pumps (see Section 8.3) the speed variation within the relevant range has

- no significant effect on the pump efficiency if the Reynolds number Re_u at nominal pump speed is higher than about $3 \cdot 10^6$
- only a minor effect on the pump efficiency if the Reynolds number Re_u at nominal pump speed is lower than about $3 \cdot 10^6$.

This minor effect is mainly caused by a small deviation from the hydrodynamic similarity law for the mechanical input power $P_{mech} \sim n^3$ and can be approximately described by a "correction factor" of the form $(n/n_{100\%})^{-\alpha}$ with a relatively small value α.

Chapter 5

Physical and Technical Background of the Electrical Power Input of Pump Units

5.1 PUMP UNITS FOR FIXED-SPEED OPERATION

An electric motor driven *single* pump unit for fixed-speed operation is schematically shown in Figure 5.1. It consists of a pump that is mechanically coupled with an electric motor fed directly from the grid ("line-fed motor"). In a *multiple* pump unit, for example a pressure boosting set, two or more single pump units are combined in a parallel arrangement and connected to common suction and discharge pipes. If all single pump units that are combined to a multiple pump unit are operated at fixed speed and are switched on or off depending on the demanded flow rate, the schematic representation in Figure 5.1 also applies for each of the individual single pump units.

The electric power input P_1 from the grid is transformed by the electric motor into the mechanical motor power P_{mech} ($=$motor output power). The latter is lower than P_1 by the total losses $\sum P_{loss,M}$ in the motor.

$$P_{mech} = P_1 - \sum P_{loss,M} \tag{5.1}$$

The mechanical motor output power P_{mech} is transmitted via the mechanical coupling[1] to the pump and hence is equal to the mechanical input power of the pump. In the pump, the mechanical input power P_{mech} is transformed into the hydraulic power P_{hyd}. The difference between P_{mech} and P_{hyd} results from the total losses $\sum P_{loss,pump}$ in the pump (described in Chapter 4).

$$P_{hyd} = P_{mech} - \sum P_{loss,pump} \tag{5.2}$$

The hydraulic power P_{hyd} is transmitted to the fluid in the hydraulic installation. According to Eq. (1.15), the hydraulic power is determined by

---

1. Only the direct mechanical coupling is considered here, which leads to the equality of motor and pump speed ($n_M = n_{pump} = n$) as well as of motor and pump shaft torque ($T_M = T_{pump} = T$). Mechanical gears or magnetic couplings are not considered.

Assessing the Energy Efficiency of Pumps and Pump Units.
DOI: http://dx.doi.org/10.1016/B978-0-08-100597-2.00005-7
© 2015 Elsevier Ltd. All rights reserved.

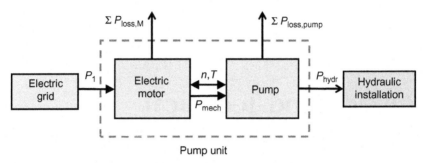

FIGURE 5.1 Schematic representation of a single pump unit for fixed-speed operation.

the flow rate Q and by the pump head H_{pump}. The pump head is directly dependent on the flow rate according to the $Q-H$ characteristic of the pump at the actual pump speed n.

Therefore, the required electric power P_1 at a certain value of the demanded flow rate Q results from

$$P_1 = \rho \cdot g \cdot Q \cdot [H(Q)]_{\text{pump}} + \sum P_{\text{loss,pump}} + \sum P_{\text{loss},M} \qquad (5.3)$$

The mechanical interaction of motor and pump is effected by the (common values of) the rotational speed n and the torque T. The exact speed n results from the respective $T-n$ characteristics of motor and pump (see below). As a first approximation, it is often assumed that the pump and motor speed is equal to the nominal pump speed $n_{\text{pump,N}}$. Under this assumption, for a certain pump

- the *pump head* is given by the $Q-H$ characteristic valid for the *nominal pump speed* $n_{\text{pump,N}}$ (as it is usually documented by the pump manufacturers in their catalogs)
- the *total pump losses* are only dependent on the flow rate Q (as discussed in Chapter 4 and exemplarily shown in Figure 4.8).

The *total motor losses* for a certain motor depend on the "load" that can be described by the motor shaft power P_{mech} or by the motor shaft torque T_M. Of course, the motor losses depend also on the type, size, and efficiency related quality of the motor.

Concerning the motor types, only the most commonly used AC induction single-speed motors will be considered here. According to Ref. [8], for these motors the total losses consist of several individual losses:

- The *electrical losses* (also called Joule losses) are caused by the electric resistance of the stator windings and in the rotor conductor bars and end rings.
- *Magnetic losses* occur in the steel laminations of the stator and rotor due to hysteresis and eddy currents.

- *Mechanical losses* result from friction in the bearings. Mostly, ventilation and windage losses are also subsumed.
- *Stray load losses* are due to various imperfections and irregularities of the electric flux within stator, rotor, and in the air gap between both.

The relative contribution of the individual motor losses to the total motor losses depends on the nominal power of the motor. According to Ref. [28], for four-pole 50 Hz motors of 0.75 kW the electrical losses amount to about 76% of the total motor losses while for 250 kW motors of the same type they contribute only by about 50% to the total motor losses. On the other hand, the relative contribution of the mechanical and stray load losses increases considerably with the nominal motor power.

Concerning the dependence of the (absolute) individual losses on the load, the electrical losses are proportional to the square of the motor current I and consequently increase extensively with the load. Also the stray load losses increase in a comparable manner with the load. The magnetic and mechanical losses are nearly constant or only weakly dependent on the load; see Ref. [8]. Therefore, the total motor losses that result as the sum of the individual losses increase in a distinct manner with the load. Within the range of variable load exerted on a line-fed motor as part of a pump unit, the dependence of the total losses (as absolute values or as losses related to the nominal motor power $P_{M,N}$) can be well approximated by a polynomial function of the second degree. This is illustrated in Figure 5.2 by the result of measurements on a 7.5 kW two-pole standard AC induction motor fed directly from the 50 Hz grid. The measurements were carried out in the frame of the project at the TU Darmstadt reported in Ref. [31].

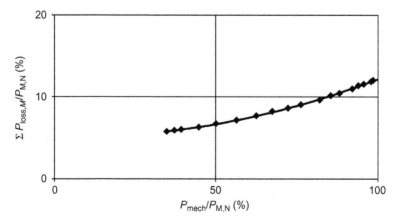

FIGURE 5.2 Related total losses $\sum P_{loss,M}/P_{M,N}$ of a 7.5 kW AC induction motor in dependence of the relative load $P_{mech}/P_{M,N}$.

The motor efficiency η_M is connected to the related total motor losses $\sum P_{loss,M}/P_{M,N}$ and to the relative load $P_{mech}/P_{M,N}$ by the relation

$$\eta_M = \frac{P_{mech}}{P_{mech} + \sum P_{loss,M}} = \frac{P_{mech}/P_{M,N}}{P_{mech}/P_{M,N} + \sum P_{loss,M}/P_{M,N}} \tag{5.4}$$

Based on the measured motor losses shown in Figure 5.2, the dependency of the motor efficiency on the relative load is calculated and presented in Figure 5.3.

Obviously, the maximum of the motor efficiency does not coincide with the nominal load but is reached at a relative load <100%. This feature is characteristic for the majority of standard AC induction motors. For these motor types, the maximum efficiency is typically obtained in the range of 60–100% of relative motor load, dependent on the motor design.

Concerning AC induction motors in the whole range of motor data that is relevant for pump units within the scope of this book, their efficiencies at the nominal operating (=full load or rated) point depend in a *fundamental* manner on the nominal motor power, on the pole number of the motor, and on the grid frequency (50 or 60 Hz). These fundamental dependencies are reflected by the equations and tables in the standard [11] for the minimum efficiency at the nominal operating point required to comply with a certain IE class. The increase of these required efficiencies with increasing nominal motor power is represented exemplarily in Figure 2.1 for a certain motor type and various IE classes. The dependency on the pole number is relatively weak, not unique in tendency and most pronounced for motors of very small size and nominal power. For equal nominal motor power, the efficiencies tend to be higher for 60 Hz than for 50 Hz.

Superposed to these fundamental tendencies of the motor efficiency at the nominal motor operating point is the efficiency related quality of the

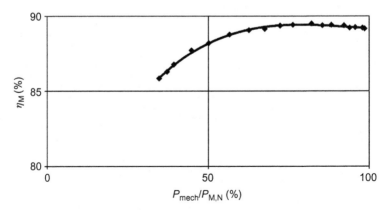

FIGURE 5.3   Motor efficiency η_M of a 7.5 kW AC induction motor in dependence of the relative load $P_{mech}/P_{M,N}$.

motors, which is reflected by the IE classes and may possibly exceed the minimum required values of an IE class. According to Ref. [8], better efficiency related motor quality can result from advanced motor design, the use of superior magnetic materials, larger copper and/or aluminum cross-sections in the stator and rotor to reduce resistance, and from other constructional and manufacturing measures.

For a pump unit consisting of the 7.5 kW motor represented by its total losses and its efficiency in Figures 5.2 and 5.3, respectively, and of an end-suction pump of appropriate nominal power the hydraulic, the mechanical and the electric power in dependence of the relative flow rate Q/Q_{BEP} are shown in Figure 5.4 as an example of measurement results of the project reported in Ref. [31]. The values of the powers are relative values related to the mechanical power P_{BEP} at the best efficiency point of the pump.

It becomes evident from Figure 5.4 that for the exemplary pump unit the total pump losses are considerably higher than the total motor losses. Qualitatively, this feature is representative for the majority of pump units.

For a more precise consideration regarding the mechanical equilibrium of pump and motor via their common values of shaft torque T and speed n, it is necessary to take into account the actual $T-n$ characteristics of both partners.

As shown in the schematic in Figure 5.5, the $T-n$ characteristic of a certain AC induction *line-fed motor* is represented by one single curve. In Figure 5.5 the torque is related to the nominal torque $T_{\text{M,N}}$ of the motor and the speed is related to the synchronous speed n_{sync} of the motor. The

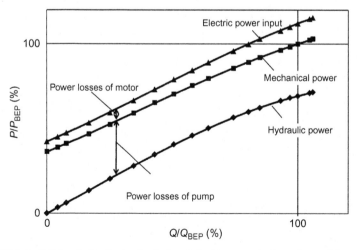

FIGURE 5.4 Relative hydraulic, mechanical, and electric power in dependence of the relative flow rate Q/Q_{BEP} for an exemplary pump unit with line-fed motor.

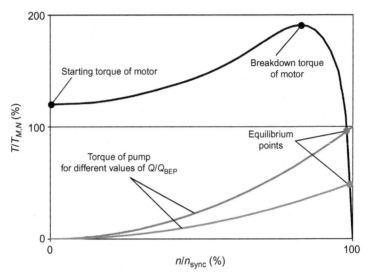

FIGURE 5.5   Torque–speed characteristics of motor and pump and equilibrium points.

synchronous speed in the unit 1/min results from the grid frequency f_{grid} and from the pole number i_{pole} of the motor:

$$n_{sync} = \frac{60 \cdot f_{grid}}{i_{pole}/2} \qquad (5.5)$$

For example, the synchronous speed is 3000 1/min for two-pole motors and 1500 1/min for four-pole motors. The $T-n$ characteristic of AC induction motors is characterized by the starting (or locked rotor) torque at zero speed, by the point of breakdown torque, and by zero torque at the synchronous speed. To the nominal motor torque $T_{M,N}$ corresponds the nominal speed $n_{M,N}$ of the motor. Because of the difference of synchronous speed and actual motor speed (the so-called rotor slip) that is necessary to generate a motor shaft torque, the motor speed decreases—with increasing torque and slip—from zero torque at $n = n_{sync}$ down to the breakdown torque. The rotor slip is usually defined as a percentage value by the equation

$$s = 100 \cdot \frac{n_{sync} - n}{n_{sync}} (\%) \qquad (5.6)$$

The nominal slip corresponding to the nominal speed, and nominal torque is usually in the order of only a few percent. It depends on the motor type, motor size, and individual motor design features.

Only the part of the $T-n$ characteristic of a motor between zero torque and breakdown torque is useable for stable operating points in combination with a pump. In general, the nominal torque of the motor is sufficiently lower than the breakdown torque, and the $T-n$ characteristic of the motor is nearly linear between synchronous speed and nominal speed. (The latter is normally slightly different from the nominal speed of the pump, which is combined with the motor in a pump unit.)

According to the hydrodynamic similarity laws, the shaft torque of a *pump* varies with the square of the speed for a constant value of the relative flow rate Q/Q_{BEP} (neglecting the possibly existing small effect of the Reynolds number as discussed in Section 4.2). This leads to $T-n$ curves corresponding to the relation

$$T = f\left(\frac{Q}{Q_{BEP}}\right) \cdot n^2 \tag{5.7}$$

that are drawn in Figure 5.5 as two exemplary red lines. The function $f(Q/Q_{BEP})$ describes the dependence of the pump torque on the (relative) flow rate at a given rotational speed, mostly determined experimentally at the nominal pump speed. At constant speed, the shape of the $T-Q$ curve of a pump is equivalent to the shape of the $P-Q$ curve of this pump, the latter usually being documented by the pump manufacturers. For the pump types and in the range of specific speeds that are relevant here, the curves show typically a decreasing mechanical power and, thereby, a decreasing shaft torque with decreasing flow rate. Under this precondition, the upper one of the red lines in Figure 5.5 corresponds to a higher value of Q/Q_{BEP} (e.g., operation at BEP at each speed) while the lower one of the red lines represents a part load operating point of the pump (at the same value $Q/Q_{BEP} < 100\%$ at each speed).

The cross-section points of the $T-n$ characteristics of pump and motor define the actual operating points, expressed by the actual values of slip, speed, and torque, resulting from the mechanical interaction of both partners. With varying relative flow rate Q/Q_{BEP} the actual speed varies slightly and deviates more or less from the nominal pump speed. Under the same precondition in respect to the function $f(Q/Q_{BEP})$ as mentioned before, the pump speed increases slightly with decreasing (relative) flow rate; see Figure 5.5. The slip and its variation with the motor load and the resulting effect on speed is more pronounced for motors of small nominal power. In fact, for pump units operated at fixed speed—or more correctly, at constant frequency of the electric power supply—the actual flow rate Q determines the actual load of the motor and, thereby, the motor losses and the motor efficiency and finally the electric power input P_1. Usually, the motor losses and the motor efficiency are determined and documented by the motor manufacturer at constant supply frequency, and the variation of motor speed and slip with the varying load is included in the resulting values and/or diagrams. This applies also for the results of measurements shown in Figures 5.2−5.4.

5.2 PUMP UNITS FOR VARIABLE-SPEED OPERATION

A *single* pump unit equipped with a Power Drive System (PDS) ($=$ motor + Complete Drive Module (CDM)^2) for variable-speed operation is schematically shown in Figure 5.6. It consists of a pump that is mechanically coupled with an electric motor fed from a CDM. In a *multiple* pump unit, all single pump units or only part of them may be operated at variable speed. Then the schematic representation in Figure 5.6 applies for those of the individual single pump units that are operated at variable speed.

In the case of pump units for variable-speed operation, the electric power input P_1 from the grid (with the fixed voltage and frequency of the grid) is first electronically converted by the CDM into the electric output power P_{el} of the CDM with a variable voltage and frequency. Because of unavoidable losses associated with the conversion, the electric output power of the CDM is lower than P_1.

$$P_{el} = P_1 - \sum P_{loss,CDM} \tag{5.8}$$

Neglecting auxiliary losses caused by the electric resistance of cables that connect the CDM and the motor and/or of electronic filters and by a power demand of necessary cooling equipment, the electric output power of the CDM equals the electric input power of the motor. The latter is transformed by the motor into the mechanical power P_{mech}. The difference between P_{el} and P_{mech} results from the total losses $\sum P_{loss,M}$ in the motor.

$$P_{mech} = P_{el} - \sum P_{loss,M} \tag{5.9}$$

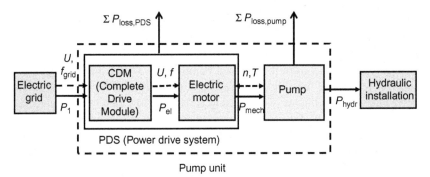

FIGURE 5.6 Schematic representation of a single pump unit for variable-speed operation.

---

2. The name CDM is more familiar in the field of electric equipment. It is used in the standard [15] and also here in the context of pump units. It has the same meaning as the often used name VSD ($=$ Variable Speed Drive).

According to Eq. (5.5) with f instead of f_{grid}, the variable output frequency f of the CDM effects a variable synchronous speed n_{sync}. The motor torque is zero at the respective synchronous speed. The actual speed n of the motor and of the directly coupled pump is slightly slower than n_{sync} because of the actual slip s as explained in Section 5.1 and illustrated schematically by Figure 5.5. The complete $T-n$ curves of the motor vary with the motor stator frequency f. Though some general characteristics of the curve shape (comparable to that one shown in Figure 5.5) are maintained, the actual $T-n$ curve at any frequency f depends also on the voltage U supplied by the CDM. This voltage varies typically with the frequency in a relation that is determined by the electronic control of the CDM. As an approximation that is often sufficiently accurate, it can be assumed that the slip at the resulting operating point is nearly constant at any frequency and, thereby, the actual motor and pump speed is proportional to the respective frequency f (this assumption is also made in the standard [15]).

The mechanical input power P_{mech} of the pump is transformed by the pump into the hydraulic power P_{hyd}. The difference between P_{mech} and P_{hyd} results from the total losses $\sum P_{loss,pump}$ in the pump (described in Chapter 4).

The required electric power P_1 at a certain value of the demanded flow rate Q results from

$$P_1 = \rho \cdot g \cdot Q \cdot [H(Q)]_{pump} + \sum P_{loss,\ pump} + \sum P_{loss,\ M}$$
$$+ \sum P_{loss,\ CDM} + \sum P_{loss,\ aux} \tag{5.10}$$

The total losses in a CDM depend in a fundamental manner on many details of its topology, of its electronic components and on operating parameters (e.g., output current and output value, switching frequency) and additionally to the efficiency related quality which is in relation to the respective IE class of the CDM (see Ref. [15] and Section 2.1). An extensive description is given in the standard [15] in connection to the mathematical modeling of the losses in CDMs. Furthermore, in the standard [15] values of the related losses[3] of defined Reference CDMs (RCDMs) are given for eight specified operating points (see Section 8.4) and for various nominal motor power values. The operating points of a CDM are defined by respective values of the frequency f and of the output current I. The latter is directly related to the motor shaft torque. It appears from the values given in Ref. [15] that the related losses of a CDM decrease with the frequency and with the output current and, thereby, with the rotational speed and with the shaft torque of the motor. Furthermore, the related losses decrease with increasing nominal power of the motor connected to the

---

3. In Ref. [15] the losses of a CDM are related to the nominal apparent power $S_{equ,N}$ that follows from the nominal output voltage $U_{out,N}$ and from the nominal output current $I_{out,N}$.

CDM. For example, the related losses of a RCDM at the operating point of nominal load[4] amount to 7.41% for $P_{M,N} = 1.1$ kW, to 3.91% for 11 kW, and to 2.74% for 110 kW. It should be noted that the losses in a CDM which is part of a pump unit are generally low compared to the losses in the pump.

A CDM combined with an AC induction motor has also an indirect effect on the total *motor losses* in the form of so-called *additional harmonic losses*. These harmonic losses result from the nonsinusoidal waveforms of voltage and current generated by the CDM and are added to the fundamental motor losses described in Section 5.1. They depend on many variables (e.g., switching frequency and control algorithm of the CDM). Various methods to determine the harmonic losses are described in the standard [13]. In the standard [15], a simplified and approximate method is proposed. According to this proposal, the total motor losses of a converter-fed motor are increased— compared to the total losses of the same motor for line-fed operation—by about 15% for motors of nominal power up to 90 kW and by about 25% for motors of nominal power higher than 90 kW. These values may be considered as nearly constant and independent of load, speed, and switching frequency provided that the switching frequency is at least 2 kHz.

According to the standard [15], for a PDS produced and sold by one manufacturer the *total losses of the PDS* as a complete product have to be determined and indicated in the product documentation. In the case a PDS is not produced and sold by one manufacturer, but components (motor, CDM, auxiliaries) produced by different manufacturers are assembled to a complete PDS, also the determination and documentation of the total losses of the complete PDS are required by the standard [15]. In both cases, the total losses of the PDS contain the losses of all components with the motor contributing also by the additional harmonic losses. Of course, the total PDS losses depend on many parameters and features of the PDS as well as on its efficiency related quality, which is in relation to the respective IES class of the PDS (see Ref. [15] and Section 2.1).

In the standard [15] values of the related losses of defined Reference PDSs (RPDSs) are given for eight specified operating points (see Section 8.4) and for a range of the nominal motor power. The operating points of a PDS are defined by the motor speed and the shaft torque, both related to their respective nominal ($=100\%$) values. The RPDSs consist each of a 400 V RCDM and a four-pole reference motor of the efficiency class IE2. Figure 5.7 based on the values given in Ref. [15] shows the related losses ($=$ related to the nominal motor power) for those three operating points that are relevant for the application of the semi-analytical model method described in Section 8.4.

---

4. In Ref. [15] the maximum load point is defined by 100% output current and, from technical reasons, by only 90% of grid frequency.

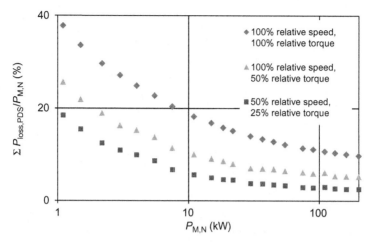

FIGURE 5.7   Related total losses $\sum P_{\text{loss,PDS}}/P_{\text{M,N}}$ of Reference Power Drive Systems (RPDSs) for three operating points in dependence of the nominal motor power $P_{\text{M,N}}$.

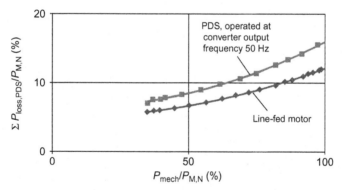

FIGURE 5.8   Related total losses $\sum P_{\text{loss}}/P_{\text{M,N}}$ of a PDS compared to those of the line-fed motor dependent on the relative motor output power $P_{\text{mech}}/P_{\text{M,N}}$.

In Figure 5.7 the distinct influence of the nominal power as well as of the operating point of a PDS appears.

To illustrate the additional losses caused by a CDM (in the CDM itself and as harmonic losses in the motor) an experimental result out of the project at the TU Darmstadt reported in Ref. [31] is presented in Figure 5.8. In this figure, the related total losses of a PDS of 7.5 kW nominal power operated at a CDM output frequency of 50 Hz are compared to the related total losses of the motor contained in the PDS when operated as line-fed motor.

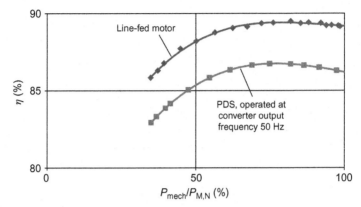

FIGURE 5.9 Efficiency η_{PDS} of a PDS compared to the efficiency η_M of the line-fed motor dependent on the relative motor output power $P_{mech}/P_{M,N}$.

It is evident that—at least at the nominal frequency $f = 50$ Hz—the total losses of the complete PDS are significantly higher than those of the motor without CDM. The ratio of both is more or less independent from the relative load that was exerted on the motor by a pump operated in the range of relative flow rates $0 \le Q/Q_{BEP} \le 100\%$. It can be expected that the result shown in Figure 5.8 is representative in a qualitative sense.

The increase of losses at nominal frequency and corresponding motor speed is also reflected by the difference in the efficiencies shown in Figure 5.9.

Regarding the electric power input P_1 of a pump unit operated at variable speed by the means of a PDS, the effect of substantially decreasing mechanical power with decreasing flow rate is the dominating one and overcompensates the slightly higher total losses of the PDS compared to the losses of a line-fed motor operated at fixed speed. This comparison is illustrated quantitatively by an exemplary result out of the project at the TU Darmstadt reported in Ref. [31]. Figure 5.10 shows the electric input power P_1 as well as the mechanical power P_{mech} in dependence of the relative flow rate Q/Q_{BEP} for the cases "variable speed" with a CDM and "fixed speed" with the same pump and the same motor, in the latter case operated as line-fed motor. In this representation, all powers related to the mechanical power at the BEP of the pump. The pump unit was operated according to a pressure control curve described by the equation

$$\frac{H}{H_{BEP}} = 0.5 + 0.5 \cdot \left(\frac{Q}{Q_{BEP}} \right) \tag{5.11}$$

The advantage of variable-speed operation in respect to the electric power input P_1 is obvious. It results from the considerably lower mechanical power

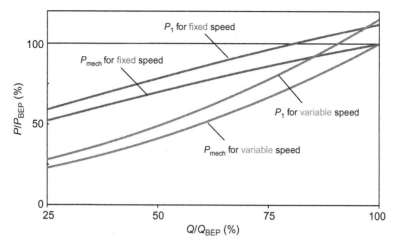

FIGURE 5.10 Comparison of the relative electric power input P_1/P_{BEP} and of the relative mechanical power P_{mech}/P_{BEP} for fixed-speed and variable-speed operation and for a pressure control curve according to Eq. (5.11).

compared to fixed-speed operation even for the pressure control curve taken as basis for the comparison. Only in a small range of relative flow rates near to the BEP, the electric power input is slightly higher for variable-speed operation caused by the additional losses of the CDM and the harmonic losses of the motor as illustrated by Figure 5.8.

Chapter 6

The Role of Manufacturing Tolerances

Electric motors, circulators, water pumps, and pump units are the focus of efficiency-related Standards and/or EU Regulations that exist or are currently being prepared or revised. They are also the focus of corresponding work within Europump. These products or extended products are typically produced in large numbers of the same type and size by the same company. For several product categories, (e.g., for electric motors and circulators), the numbers produced per year are very large. These products are really mass-produced. An entirety of electric devices (motors, complete drive modules (CDMs), power drive systems (PDSs)), pumps, or circulators, each of the same type and size and manufactured in series by the same manufacturer, are here simply called a size. If complete pump units of the same configuration are placed on the market or put into service by the same company in quite large numbers, this entirety of pump units is called type series.

In these cases, the classification in the form of efficiency classes or efficiency indicators is representative for the entirety of all individual products or extended products belonging to the same size or type series.

Only in those (rather rare) cases that an extended product (= a complete pump unit) is placed on the market or put into service as a unique one or in a very limited number, the classification may be related to the individual subjects.

Regarding the classification of a whole size or type series, the role and effect of *manufacturing tolerances* have to be considered.

Within such an entirety, there exists an inevitable scatter of the performance characteristics of the individual specimens including the efficiency and other values that are of relevance for the efficiency indicators as the Minimum Efficiency Index (MEI) or Energy Efficiency Index (EEI) (see Chapters 7 and 8).

The scatter of performance characteristics results from small differences of physical quantities that are relevant for losses and efficiency. These quantities are

- for pumps and for the hydraulic part of circulators
  - geometrical dimensions (as, e.g., internal flow cross sections, impeller blade angles, gap clearances)
  - the effective roughness of wetted surfaces

  both influenced by manufacturing procedures

Assessing the Energy Efficiency of Pumps and Pump Units.
DOI: http://dx.doi.org/10.1016/B978-0-08-100597-2.00006-9
© 2015 Elsevier Ltd. All rights reserved.

- for electric components in general properties of raw materials (particularly copper and magnetic steel)
- for motors in particular also geometrical dimensions influenced by manufacturing procedures.

The scatter of these geometrical and material quantities (within the range of allowable tolerances fixed by the respective manufacturers) is inherent in every production process and cannot be reduced below some economically acceptable limits.

The bandwidth of scatter of the relevant geometrical and material quantities leads to a corresponding bandwidth of performance characteristics of individual specimens of products/extended products within the same size/type series. The maximum differences of performance characteristics within the same entirety caused by the production process are usually called *manufacturing performance tolerances*.

A different magnitude of losses caused by the production process affects directly the efficiency at the nominal operating point of pumps and electric motors and the numerical value of efficiency-related indicators. This effect shall be illustrated by the following examples.

The relation between the total related losses and the efficiency of a pump is described by Eq. (4.2) in Section 4.1. Per definition, the pump losses are related to the nominal *input* power P_{BEP}. Especially for the best efficiency point (BEP = nominal operating point of pumps) it is

$$\eta_{pump,\text{BEP}} = 1 - \frac{\sum (P_{loss,pump})_{\text{BEP}}}{P_{\text{BEP}}} \tag{6.1}$$

In Eq. (6.1), the efficiency and the related losses are decimal numbers[1] <1.0.

Based on this relation, a difference of the related pump losses at BEP causes a difference of the pump efficiency at BEP. For example, an increase of the related pump losses at BEP by a factor 1.1 causes a reduction of the pump efficiency $\eta_{pump,\text{BEP}}$ from 80.0% to 78.0%, or from 60.0% to 56.0%.

The losses of electric devices are—per definition—related to the nominal *output* power. Therefore, the relation between the total related losses and the efficiency of electric motors at the nominal operating point N is described by

$$\eta_{M,N} = \frac{1}{1 + \frac{\left(\sum P_{loss,M}\right)_N}{P_{M,N}}} \tag{6.2}$$

---

1. Multiplying the value of the efficiency in Eq. (6.1) by the factor 100 yields the pump efficiency in [%] as it is more usually stated.

In Eq. (6.2) the efficiency and the related losses are decimal numbers.[2] Based on this relation, a difference of the related motor losses at nominal load causes a difference of the motor efficiency at nominal load. For example, an increase of the related motor losses at nominal load by a factor 1.1 causes a reduction of the motor efficiency $\eta_{M,N}$ from 95.0% to 94.5%, or from 80.0% to 78.4%.

Concerning the effect of differences in losses on the EEI, not only the losses at the nominal operation point but also at all load points that are relevant for the determination of EEI (see Section 8.1) must be taken into account. To illustrate the effect, an example was calculated by application of the semi-analytical model (SAM) of pump units (see Section 8.4). The results apply to the same pump unit that served as an example for the results shown in Figure 5.10. The pump efficiency at BEP is 63.8%, and the efficiency of the PDS at its nominal load point is 85.3%. The reference flow-time profile for variable flow systems (see Section 8.1) and the reference pressure control curve according to Eq. (5.11) were chosen for the calculations. In this example, an increase of the related *pump* losses at *all* relevant hydraulic load points by a factor 1.1 causes an increase of the value of EEI by 4.4%. On the other hand, an increase of the related losses of the *PDS* at *all* relevant mechanical load points by a factor 1.1 causes an increase of the value of EEI by 1.9%.

A certain size of pumps or circulators or a certain type series of pump units (in the sense explained earlier) is characterized by its *mean performance data and curves* that are usually presented in catalogs or other documents. In a strong sense, these mean values are defined by

$$x_{mean} = \frac{1}{z} \cdot \sum_{i=1}^{z} x_i \qquad (6.3)$$

In Eq. (6.3), x is the respective performance quantity (for example Q, H, η_{pump}, η_M, P_1), i is the consecutive number, and z is the total number of individual specimens of the same size or type series ever produced and placed on the market and/or put into service. For sizes or type series produced in very large numbers, the total number z can be replaced by the number z_T of individual products/extended products of the same size/type series produced in a sufficiently long time period T.

If the number z (or z_T) is sufficiently large, it can generally be assumed that the geometrical dimensions and material properties and, thereby, the resulting performance quantities x that are influenced by them show so-called *normal or Gaussian distributions* (Figure 6.1).

---

2. Multiplying the value of the efficiency in Eq. (6.2) by the factor 100 yields the motor efficiency in [%] as it is more usually stated.

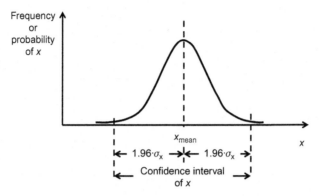

FIGURE 6.1   Normal or Gaussian distribution of a quantity x.

These distributions are characterized by their mean value x_{mean} (which is identical to the value defined by Eq. (6.3)) and by the associated bandwidth of scatter around the mean value. The normal distribution describes the probability of individual values x within their whole range. The "width" of the distribution depends on the standard deviation σ_x of the values of the quantity x:

$$\sigma_x = \sqrt{\frac{1}{z-1} \cdot \sum_{i=1}^{z} (x_i - x_{mean})^2} \qquad (6.4)$$

For normal distributions, the probability is 95% that any value of the respective performance quantity x is confined to the so-called *confidence interval*. The confidence interval has the width $\pm 1.96 \cdot \sigma_x$ around the mean value x_{mean} and can serve to quantify the manufacturing tolerance $t_{man,x}$ of a performance quantity x.

Following from the general definitions of the normal distribution and of the standard deviation, the individual value x has a probability of

- 50% each to be higher or lower than x_{mean}
- 97.5% to be equal to or higher than $x_{mean} - 1.96\ \sigma_x$
- 97.5% to be equal to or lower than $x_{mean} + 1.96\ \sigma_x$.

However, the value can be lower than $x_{mean} - 1.96\ \sigma_x$ or higher than $x_{mean} + 1.96\ \sigma_x$, but in both cases only with a probability of 2.5%.

The true mean value of a performance quantity x, especially of the efficiency or of an efficiency indicator, as well as the corresponding manufacturing performance tolerance $t_{man,x}$ could only be determined exactly by Eqs (6.3) and (6.4), respectively, if the values x_i of all z or z_T individual specimens of the respective size or type series were known (e.g., from tests). Since in most cases the total numbers z or z_T are too large, it is normally not possible for economic reasons to test each individual specimen.

Therefore, only estimates of the true mean values and of the corresponding confidence interval (or manufacturing tolerance) can be determined

- either on the basis of tests on a sample consisting of a limited number (that is typically very small compared to z or z_T) of specimens drawn randomly out of the size or type series
- or—in the case of pump units—by the application of SAMs (see Section 8.4).

It is important to note that—on the basis of tests on a sample consisting of a limited number of specimens—the true mean values of the efficiency or of an efficiency indicator can only be determined to be confined to the confidence interval with a probability of 95%.

Besides the effect and consequences of manufacturing tolerances on values of efficiency or efficiency indicators, also the measurement uncertainties (see Sections 8.3 and 8.4) have to be taken into account in respect to values indicated on nameplates or in documents by the responsible producers.

Recommended methods for the determination of MEI_{mean} of a pump size and of EEI_{mean} of a type series of pump units by sample tests will be provided in Ref. [18] and in a standard on EEI of pump units currently in preparation, respectively.

For electric motors, the Standard IEC 60034-1 [41] defines maximum allowable deviations of the efficiency (at nominal load) resulting from a test on any individual motor from the value declared on the nameplate or in the catalog. This tolerance is explained in Ref. [41] to cover variations in raw material properties and manufacturing procedures but not to cover measurement uncertainties. It is the manufacturing tolerance of motor efficiency. The values of the manufacturing tolerance defined in Ref. [41] are -15% of $(1 - \eta_{M,N})$ for motors up to and including 150 kW and -10% of $(1 - \eta_{M,N})$ for motors above 150 kW. (In this context, $\eta_{M,N}$ is specified as a decimal number <1.0.)

Concerning the energy efficiency class IESx of a PDS or IEx of a CDM, the Standard EN 50598 [15] specifies that the responsible manufacturer shall consider the uncertainty of the used loss determination method (test or application of mathematical models). This uncertainty shall be added to the determined loss value of the PDS or CDM, and the corrected value of power losses shall apply for the energy efficiency classification (see Section 2.1) and shall be used to define related power losses of the PDS or CDM, respectively. No direct information on manufacturing tolerances of related losses is given in Ref. [15].

Of course, the probability of the declared values to be confirmed by a *verification* procedure in the frame of market surveillance should be as high as possible. Regarding the declaration of efficiency values or efficiency indicators for pumps or pump units, recommendations in this respect are (or will be) also provided in Ref. [18] and in a standard on EEI of pump units currently in preparation, respectively.

The verification procedure described in EU regulations [20,22,24] consists in general (with only few exceptions) of the following steps:

1. Test of one single specimen according to referenced test standards.
2. The corresponding size is considered to comply with the provisions set out in the regulation if the respective requirements on the efficiency or on the efficiency indicator are achieved. Achievement takes account of a "verification tolerance" (=maximum accepted deviation of the value found out in the verification test from the declared value).
3. If the result referred to in point 2 is not achieved, three additional specimens are drawn randomly out of the same size.
4. The corresponding size is considered to comply with the provisions set out in the regulation, if in the average the respective requirements on the efficiency or on the efficiency indicator are achieved—as described in point 1—by the three additional specimens.
5. If the results referred to in point 4 are not achieved, the size shall be considered not to comply with the respective regulation.

The verification tolerance mentioned in step 2 is differently defined in the three EU regulations.

According to the EU regulation for electric motors [20], the tolerance applies to the motor losses $(1 - \eta_{M,N})$ at nominal load. Compliance is achieved if these losses resulting from the verification test(s) do not vary from the values according to the minimum required efficiency for the respective IE class by more than 15% in the power range from 0.75 to 150 kW and 10% in the power range >150 up to 375 kW.

According to the EU regulation for circulators [22], the tolerance is applied to the EEI value. Compliance is achieved if the values of EEI resulting from the verification test(s) do not exceed the values declared by the manufacturer by more than 7%. (For the definition and determination of the EEI of circulators, see Section 8.2.)

According to the EU regulation for water pumps [24], compliance is achieved if the pump efficiency measured in a verification test at each of the conditions BEP, PL, and OL (η_{BEP}, η_{PL}, η_{OL}) does not vary below the values corresponding to the respective value of MEI by more than 5% (=percentage of the η-values). (For the definition of MEI and the relation to minimum required efficiencies of water pumps, see Chapter 7.)

The values of efficiencies or efficiency indicators that result from the application of the verification tolerance to the values that are specified by standards or regulations and that are declared by manufacturers can be considered as threshold values of the verification procedure. It is evident that the probability for a size to comply with the respective EU regulation is highest

- if all possible values of the efficiencies η of a size covered by the manufacturing performance tolerances are *at least equal to or higher* than the respective threshold values

- if all possible values of EEI of a size or type series covered by the manufacturing performance tolerances are *at least equal to or lower* than the respective threshold value.

The smaller the manufacturing performance tolerances $t_{\mathrm{man},\eta}$ or $t_{\mathrm{man,EEI}}$, respectively, are, the more easily these conditions can be met by a size or by a type series.

Chapter 7

The Concept of the Minimum Efficiency Index (MEI) for Pumps

The concept of the Minimum Efficiency Index (MEI) was originally proposed by Europump and developed by the TUD [30] in the frame of the product approach for pumps (see Chapter 3). It was then implemented into the EU Regulation [24] and in the corresponding standard [18] for the application on clean water pumps (see Chapter 2).

The basic principles of the concept of MEI are the following:

- MEI serves as an indicator for the efficiency-related quality of pumps.
- The definition of MEI incorporates and compensates fundamental dependencies of pump efficiency on the pump type and on relevant nominal pump data (see Chapter 4). Therefore, MEI serves to quantify and assess the efficiency-related quality of pumps independently of individual pump attributes.
- There is a direct correlation between numerical values of MEI and corresponding minimum required (mean) efficiency values of a size.[1] Required minimum MEI values correspond to required minimum efficiency values of a size.
- Besides the pump efficiency at best efficiency point (BEP = nominal operating point), also the pump efficiencies at moderate part-load and overload operation are incorporated in the definition of MEI.
- Minimum required MEI values stated by legislation determine values of (mean) efficiency to be attained or exceeded by a size for conformity.
- Actual MEI values of a size that are better than those required by legislation can bring advantages for manufacturers and end users on the market.

Concerning the incorporation of the efficiencies at two additional operating points besides BEP, account is taken of two facts:

- Typically, the product program of pump manufacturers for a certain pump type is subdivided into a limited number of different pump sizes,

---

1. The word "size" is used here (as explained in Chapter 6) for an entity of pumps of the same type, of the same nominal data, and produced by the same manufacturer.

Assessing the Energy Efficiency of Pumps and Pump Units.
DOI: http://dx.doi.org/10.1016/B978-0-08-100597-2.00007-0
© 2015 Elsevier Ltd. All rights reserved.

each covering a certain range of flow rate Q and pump head H. Not for any $Q-H$ duty point specified by a customer will a pump size exist with a BEP that coincides with the specified duty point. Therefore, the selection of a pump out of the available sizes will most often cause the duty point to be a slight "off-design" point of the selected pump, and the efficiency at this point is equally important as that one at BEP.

- As already commented in Section 4.2, the great majority of pumps is not operated exclusively at one single operating point (expressed by the relative flow rate Q/Q_{BEP}) but is operated within a range of operating points. This applies also for pumps operated at fixed speed in pumping systems of the constant flow type (see Sections 1.2 and 8.1). The variation of the operating point can result from variations of the hydraulic resistance of the installation (caused either by adjusting a varying demand of flow rate by a control valve or by long-time effects, such as internal incrustation of pipes) or, in the case of parallel operation of pumps, from variable operation conditions when different numbers of pumps are running.

7.1 DEFINITION OF MEI

According to the concept and definition of MEI, a size is only qualified for a certain value of MEI if the following three conditions at the operating points denoted as BEP, PL (part-load), and OL (overload) are met:

$$\eta_{BEP,\ mean} \geq \eta_{BEP,\ min\ requ} \tag{7.1}$$

$$\eta_{PL,\ mean} \geq \eta_{PL,\ min\ requ} \tag{7.2}$$

$$\eta_{OL,\ mean} \geq \eta_{OL,\ min\ requ} \tag{7.3}$$

The operating points that shall be representative for the efficiency in the PL and OL range are fixed by definition at $Q_{PL} = 75\%$ of Q_{BEP} and $Q_{OL} = 110\%$ of Q_{BEP}.

All efficiency values in the conditions given in Eqs (7.1)–(7.3) are mean values of the size and apply for pumps of this size with *full impeller diameter*. The full diameter is the maximum outer impeller diameter for which performance characteristics of a pump size are presented in the documentation (e.g., catalogs) of a manufacturer.

However, it is important to note that for many applications, the duty point specified by the customer can only be achieved by selecting the most appropriate pump with a trimmed impeller. In spite of the fact that trimming the impeller normally reduces the pump efficiency, the use of pumps with a trimmed impeller can be more favorable in respect to the energy efficiency of the whole pumping system compared to using the pump with its full diameter and throttling the "unneeded" part of the pump head. This is especially valid for pumps operated at fixed speed.

A necessary and important task was to define quantitatively the minimum required efficiencies in dependence on the numerical value of MEI. These definitions and relations were elaborated in a project at the TUD [30] and agreed on by the responsible Joint Working Group of Europump. The results are based on scientific analyses of the attainable efficiency of pumps [35,37,39] and particularly on the statistical evaluation of data collected from several questionnaires sent to European pump manufacturers in 2007.

The collected data comprised all pump types within the (identical) scope of the EU Regulation [24] and of the standard [18] for clean water pumps (see Chapter 2). Performance data for nominal speed values according to operation in combination with two-pole as well as with four-pole AC induction motors were collected. The specific speed n_s of the pump sizes in the data base ranges from 6 to 110.5 1/min and the range of Q_{BEP} is from 1.8 to 1200 m³/h. The performance data supplied by the European manufacturers were valid for the full diameter of the respective pump sizes. Altogether, the collection consisted of about 2400 data sets.

For the minimum efficiency $\eta_{BEP,min\ requ}$ at BEP, the following mathematical correlation was derived that reflects the fundamental dependencies of η_{BEP} on the flow rate Q_{BEP} and on the specific speed n_s explained in Section 4.1:

$$\eta_{BEP,\ min\ requ} = -11.48 \cdot (\ln(n_s))^2 - 0.85 \cdot (\ln(Q_{BEP}))^2 - 0.38 \cdot \ln(n_s) \cdot \ln(Q_{BEP})$$
$$+88.59 \cdot \ln(n_s) + 13.46 \cdot \ln(Q_{BEP}) - C$$

$$(7.4)$$

From this equation the efficiency $\eta_{BEP,min\ requ}$ results in (%). The specific speed has the unit 1/min, and the flow rate Q_{BEP} has the unit m³/h. Q_{BEP} and n_s appear in Eq. (7.4) in the form of their natural logarithms. The efficiency calculated by Eq. (7.4) shall be rounded to the first decimal.

Eq. (7.4) applies to all pump types and nominal speeds within the (identical) scope of the EU Regulation [24] and of the standard [18]. The constant C (in [%]) depends on MEI and additionally on the type and nominal speed of the size.

The mathematical validity of the equation is limited to the ranges $6 \leq n_s \leq 120$ 1/min and $2 \leq Q_{BEP} \leq 1000$ m³/h. The physical range of efficiency values calculated by Eq. (7.4) is limited to $\eta_{BEP,min\ requ} \leq 88\%$. The latter limitation is caused by the facts that the losses in commercially designed and manufactured pumps cannot be reduced below a lower limit without unacceptable efforts and costs and/or without incompatibility with other requirements mentioned in Section 4.1.

Generally, the equation is valid for the *full impeller diameter* of a pump size. In the case of multistage pumps (MS), Eq. (7.4) is valid for a minimum stage number of $i_{st} = 3$, in the case of submersible multistage pumps (MSS) for a minimum stage number of $i_{st} = 9$.

In the project reported in Ref. [30], the pump data given by the manufacturers was also evaluated at 75% and at 110% of Q_{BEP}. The ratios of η_{PL}/η_{BEP} and η_{OL}/η_{BEP} were calculated. As a result, the following equations were stated to be appropriate for the concept and definition of MEI and shall be applied to all types and nominal data of pumps in the scope of Refs [18,24]:

$$\eta_{PL, \text{ min requ}} = 0.947 \cdot \eta_{BEP, \text{ min requ}} \tag{7.5}$$

$$\eta_{OL, \text{ min requ}} = 0.985 \cdot \eta_{BEP, \text{ min requ}} \tag{7.6}$$

It is favorable for a high value of MEI if the mean efficiency curve of a size shows a *high maximum* and is as *flat as* possible in the range from Q_{PL} to Q_{OL}.

The values of the constant C in Eq. (7.4) were determined in Ref. [30] by the condition that the value of MEI, multiplied by 100, is equivalent to the percentage of pump sizes that did not (at the time of the data collection in 2007) meet the conditions according to Eqs (7.1)−(7.3) (=so-called "cut-off-effect" of the MEI values that are required by legislation). The numerical values of the constant C are presented in a table (identical) in Refs [18,24] for the relevant pump types and nominal speeds and for various values of MEI in steps of 0.1.

At the lower limit MEI = 0, the corresponding minimum required efficiencies can be achieved on a low level of design and manufacturing. For values of MEI > 0.7 the corresponding efficiencies can only be achieved by a very special hydraulic design without respecting other design aspects (see Section 4.1) and by extraordinary measures in mechanical design and manufacturing. Therefore, values of MEI higher than 0.7 are not practically attainable for mass-produced pumps.

To illustrate the relation between MEI and the minimum required efficiency at BEP, the latter is calculated exemplarily by Eq. (7.4) with the values of C taken from the table in Refs [18,24] for three different pump types and for certain values of nominal speed n_N, flow rate Q_{BEP}, and specific speed n_s. The calculated values are shown in Figure 7.1.

The obvious differences of minimum required efficiency at BEP for different pump types at equal nominal data can be explained by the following reasons:

- Compared to end suction pumps (represented by the type ESOB in Figure 7.1), the efficiencies of multistage pumps (type MS) are lower mainly because of additional through-flow losses in the guide vanes and return channels.

- For pumps of the type ESCCi (that are close-coupled to the motor and have no own bearings) the mechanical friction losses are slightly lower compared to pump types with own bearings. The all in all higher total losses and lower efficiencies of the pump type ESCCi compared to

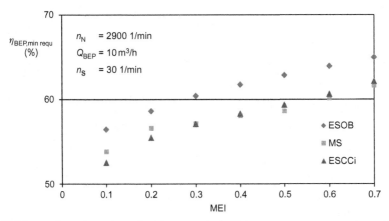

FIGURE 7.1 Exemplary values of minimum required efficiency $\eta_{BEP,min\ requ}$ at certain nominal pump data for various pump types and values of MEI.

non-inline pumps, though, result from the dominating effect of through-flow losses (see Section 4.1). These are higher for inline pumps because of the special casing geometry that is necessary to achieve the inline condition for casing inlet and outlet.

Further information is contained in Figure 7.1 The presented exemplary values make clear the effect (and its order of magnitude) of an increase of the minimum MEI from 0.1 to 0.4 (required by EU Regulation [24] for the date January 1, 2015) on the minimum required efficiencies of pump sizes.

7.2 TEST PROCEDURES

The determination of the efficiency values at the three relevant operating points BEP, PL, and OL is based on (sample) tests and evaluations of measured data. The method and requirements on test conditions and maximum permissible measurement uncertainties (with reference to the ISO Standard for acceptance tests [17], class 2) are described in detail in the standard [18]. The determination of the values Q_{BEP} and η_{BEP} is carried out in the way described in Section 4.1 and illustrated by Figure 4.1. The values η_{PL} and η_{OL} result from the best-fit $Q-\eta$ curve determined on the basis of the test points. For the determination of the *mean* efficiency values of a pump size by tests and evaluations

- either on only one individual pump out of a size
- or on a sample consisting of a limited number of pumps out of the same size

recommendations are given in an Annex of the standard [18].

Because of the unavoidable *measurement uncertainties* of all quantities (Q, H, P_{mech}, n) that need to be measured and of the effect of error propagation in the evaluation procedure, the values of Q_{BEP} and of the efficiencies η_{BEP}, η_{PL}, and η_{OL} of a tested *individual* pump can only be determined to be confined to a respective confidence interval of 95% probability. Usually, the magnitude of these confidence intervals is set to about twice the (absolute) value of the total measurement uncertainties $e_{\text{tot,x}}$ of the quantities x that result from test and evaluation. The total measurement uncertainties and the corresponding magnitude of the confidence intervals depend on

- the *systematic or instrument uncertainty* $e_{\text{s,x}}$ which can be reduced by measures such as use of highly accurate instrumentation and data acquisition systems, careful installation of measuring devices, and careful calibration of instruments
- the *random uncertainty* $e_{\text{r,x}}$ that can be reduced by increasing the number of instrument readings for the same test point and measured quantity, by providing damping devices in the measuring chains and/or—in the case of using electronic data acquisition systems—by appropriate low pass filtering, data sampling, and averaging.

Under the assumption that $e_{\text{s,x}}$ and $e_{\text{r,x}}$ are statistically independent of each other, the total measurement uncertainty results from

$$e_{\text{tot,x}} = \sqrt{e_{\text{s,x}}^2 + e_{\text{r,x}}^2} \tag{7.7}$$

Within the confidence intervals, every value of x of the tested individual pump is equally valid.

It is important to note that—in respect to MEI and the corresponding minimum required efficiencies—also the specific speed n_{s} is relevant (see Eq. (7.4)). Besides the nominal speed n_{N} and the flow rate Q_{BEP}, the definition of the specific speed (see Eq. (4.20)) contains also the pump head H_{BEP}. Therefore, in addition to Q_{BEP} the corresponding value H_{BEP} has to be determined by test and evaluation. From a best-fit (polynomial) function for the measured Q–H points, the value of H_{BEP} at $Q = Q_{\text{BEP}}$ can be easily calculated. Also the value of n_{s} of a tested individual pump can only be determined to be confined to the respective confidence interval.

According to Chapter 6, all (true) values $x_{i,\text{true}}$ of one tested individual pump out of a size can be expected to be within the respective interval $x_{\text{mean}} \pm t_{\text{man,x}}$ with a probability of 95%. Vice versa, the (unknown) value of x_{mean} can be expected to be (with a probability of 95%) within the interval $x_{i,\text{true}} \pm t_{\text{man,x}}$ around the value $x_{i,\text{true}}$ of an individual pump. Because of the effect of measurement uncertainties explained before, also the values $x_{i,\text{true}}$ of a tested individual pump cannot be determined exactly, but only to be confined with a probability of 95% to the respective interval $x_{i,\text{test}} \pm e_{\text{tot,x}}$. The manufacturing performance tolerances $t_{\text{man,x}}$ and the total measurement uncertainties $e_{\text{tot,x}}$ of quantities x (Q_{BEP}, η_{BEP}, η_{PL}, η_{OL}, n_s) that are relevant for the assessment of a size in respect to MEI can be assumed to be statistically independent of each other. Therefore, based on a test on an individual pump the mean values x_{mean} of a quantity x can be expected to be confined to the interval $x_{i,\text{test}} \pm t_{\text{tot,x}}$ with the total tolerance (comprising manufacturing tolerance *and* measurement uncertainty)

$$t_{\text{tot,x}} = \sqrt{t^2_{\text{man,x}} + e^2_{\text{tot,x}}} \qquad (7.8)$$

If no reliable information exists on the magnitude of the manufacturing tolerances $t_{\text{man,x}}$ of the relevant quantities x, estimations or default values must be used. Recommendations are given in an Annex of the standard [18]. Measures to reduce the maximum possible deviation of x_{mean} from x_{test} are

- reduction of the actual manufacturing tolerances by more precise manufacturing technologies and/or processes
- reduction of the measurement uncertainties
- testing a sample consisting of N individual pumps out of the same size (with preferably $N \geq 5$) and application of well-established statistical methods on the results.

The statistical evaluation of sample tests is also described in an Annex of the standard [18].

Finally, it is the responsibility of the pump manufacturer to determine and state the mean values of Q_{BEP}, η_{BEP}, η_{PL}, η_{OL}, and n_s to be used for the assessment of a pump size by the manufacturer (so-called qualification) in respect to MEI. The stating of the values used for the qualification should secure a high probability of being approved by a verification procedure (see Chapter 6).

7.3 ASSESSMENT

To qualify a pump size for a minimum value required by the EU Regulation [24] (or a future revision) and, thereby, for conformity with the EU Regulation, the mean values of efficiency must meet all three conditions described by Eqs (7.1)–(7.3). This is illustrated schematically in Figure 7.2. In the case shown in Figure 7.2, the conditions are fulfilled at all three relevant operating points PL, BEP, and OL with some margins, the smallest margin existing at the point PL.

It has to be noted that in the *verification* procedure described in Chapter 6, in contrast to Figure 7.2, the efficiency values resulting from the test(s)—instead of the mean values determined by the manufacturer—are compared to the threshold values that result from the minimum required values multiplied by the factor 0.95, which corresponds to the verification tolerance of 5%.

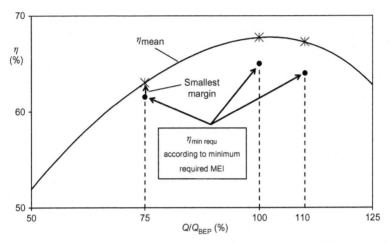

FIGURE 7.2   Qualification of a pump size for a minimum required value of MEI (schematic).

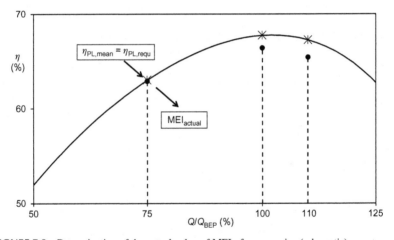

FIGURE 7.3   Determination of the actual value of MEI of a pump size (schematic).

The determination of the actual numerical value of MEI of a pump size by the manufacturer is illustrated schematically in Figure 7.3.

In this case, in a first step one of the three relevant operating points has to be selected, which is representative in respect to MEI. In the example shown in Figure 7.3, this is the operating point PL. However, for other sizes the operating points BEP or OL can be representative for MEI. To find out the representative operating point, Eq. (7.9) (the "reverse version" of Eq. (7.4)) is applied on all three relevant operating points PL, BEP, and OL and that one is selected, which leads to the greatest value of C.

$$C = -11.48 \cdot (\ln(n_s))^2 - 0.85 \cdot (\ln(Q_{BEP}))^2 - 0.38 \cdot \ln(n_s) \cdot \ln(Q_{BEP})$$
$$+88.59 \cdot \ln(n_s) + 13.46 \cdot \ln(Q_{BEP}) - \eta_{mean} \tag{7.9}$$

Then, by linear interpolation between the discrete MEI values given in the table in Refs [18,24], the numerical value of MEI corresponding to the value of C of the selected operating point is calculated and rounded to the second decimal.

Chapter 8

The Concept of the Energy Efficiency Index (EEI) for Circulators and Pump Units

8.1 GENERAL ASPECTS AND DEFINITIONS

8.1.1 Reasons for the Introduction of EEI

The efficiency related classification of electric devices (IE- and IES-classes) applicable for the drive part of pump units (motors, power drive systems (PDSs), complete drive modules (CDMs)) is based—according to the existing relevant standards [11,15]—exclusively on the device efficiency at the nominal operating condition (=nominal load). The efficiency related quality of clean water pumps is quantitatively assessed—according to the EU Regulation [24] and the standard [18]—by their Minimum Efficiency Index (MEI) value, which is based on efficiencies in a relatively limited range of operating points (=relative flow rates Q/Q_{BEP}).

The assessment of energy efficiency of configurations consisting of pumps and their drives only on the basis of the classification of their components by respective IE-, IES-, or MEI-values would not take account of different modes of operation that are typical for pumping systems (see Section 1.2 and the following).

For example, in many applications the demanded flow rate and the corresponding operating points of the pumps and their drives vary in a wide range. As a consequence, the efficiencies of pumps and drives at low part-load are at least equally or even more important for the energy consumption than the efficiencies at the nominal condition.

As explained in Section 1.2 and described by Eq. (1.21), the actual energy consumption in a certain time period of operation depends not only on the pump and drive efficiencies within the range of operating points but also on the flow-time profile, the corresponding values of demanded head, and the kind of flow rate adjustment (throttling or speed variation).

Assessing the Energy Efficiency of Pumps and Pump Units.
DOI: http://dx.doi.org/10.1016/B978-0-08-100597-2.00008-2
© 2015 Elsevier Ltd. All rights reserved.

From these reasons, a meaningful indicator of the energy efficiency of configurations of pumps and drives should include operation apart from the nominal point, should take into account the mode of operation, and should reflect the actual energy consumption, not only the pump and drive efficiencies.

An increasing amount of energy saving can be achieved more effectively by requiring more ambitious values of an appropriately defined indicator (as the Energy Efficiency Index (EEI); see below) than by tightening the requirements on the MEI-value of pumps and/or on the efficiency class of the pump drive. Better values of the indicator can result from

- selection of components due to better part-load efficiencies
- realization of the most adequate mode of operation and kind of flow rate adjustment
- "intelligent" design and control in the case of multiple pump units.

In the case of multiple pump units, speed variation of all pumps contained in the unit does not necessarily minimize the energy consumption.

An appropriate energy efficiency related indicator for configurations consisting of pumps and drives is the EEI. It is well established for assessing the energy efficiency of circulators, and it is now proposed for further applications in the field of pumps.

8.1.2   Definition of EEI

The EEI is a dimensionless value. It is defined by the following equation:

$$\text{EEI} = \frac{P_{1,\text{avg}}}{P_{1,\text{ref}}} \tag{8.1}$$

In this definition, $P_{1,\text{avg}}$ is the weighted average of the electric power input P_1. It is calculated by the equation

$$P_{1,\text{avg}} = \sum_{i=1}^{i=N} \left[\left(\frac{\Delta t}{t_{\text{tot}}} \right)_i \cdot P_{1,i} \right] \tag{8.2}$$

In Eq. (8.2) i is the consecutive number of the points of a *reference flow-time profile* and N is the total number of the points of this *flow*-time profile (see Eq. (1.2) in Section 1.2). Reference *flow*-time profiles are defined as described below.

The denominator of Eq. (8.1) is the reference electric power $P_{1,\text{ref}}$. This reference power shall be defined in such a way that the resulting value of EEI reflects directly the energy efficiency related quality. Therefore, the value of $P_{1,\text{ref}}$ shall compensate the fundamental influences on the efficiency of the pump and its drive, such as their geometrical size, nominal speed, and nominal power. For this purpose, an appropriate definition of the reference power $P_{1,\text{ref}}$ shall

- have a physical basis
- take into account relevant influences on the electric power input of configurations consisting of pumps and drives
- be independent of the efficiency related quality of individual components of the configuration.

The precise definition of $P_{1,\text{ref}}$ that meets these requirements can be different for different categories of configurations of pumps and drives.

With only a few exceptions (to be specially justified), the values $P_{1,\text{avg}}$ as well as $P_{1,\text{ref}}$ have—per definition—to be determined and inserted in Eq. (8.1) for the *full diameter* of the pump(s) that is (are) part of the configuration.

According to the definition of EEI, the operation of complete configurations of pump(s) and drive(s) is more energy efficient the *lower* the numerical value of EEI is.

8.1.3   Reference Flow-Time Profiles

The reference flow-time profiles that are needed to calculate the average electric input power $P_{1,\text{avg}}$ (the numerator of EEI) are defined in relation to the type of the system that is supplied with the pumped liquid and to the mode of operation of the pump.

In respect to the demanded flow rate, two characteristic types of systems can be distinguished:

- systems with demanded flow rates varying only slightly around a 100% value
- systems with demanded flow rates varying in a wide range below a 100% value.

For simplification of the wording, the following expressions are used for the corresponding reference flow-time profiles:

- "constant flow systems" for systems with slightly varying demand of flow rate
- "variable flow systems" for systems with widely varying demand of flow rate.

The range of demanded flow rate is determined by the process that is served by the respective installation.

In constant flow systems, the slight variations of the flow rate around the nominal value are most often caused by secondary influences from the process, such as due to the varying level of liquid in reservoirs. The variation of flow rate occurs typically within the range that is covered by the definition of the MEI of pumps (see Refs [18,24] and Chapter 7) and which is from 75% to 110% of $Q_{100\%}$.

In variable flow systems, the demanded flow rate is typically much lower than $Q_{100\%}$ for considerable fractions of the total operating time. The demand of flow rate varies typically within a range from 25% of $Q_{100\%}$ to $Q_{100\%}$. (In the special case of booster stations for buildings, the lower limit of demanded flow rate is typically as low as 10% of $Q_{100\%}$, see Section 8.3.)

In respect to the rotational speed of the pump, two different modes of operation exist. Configurations consisting of pumps and line-fed AC induction motors can only be operated at constant motor stator frequency and thereby (neglecting the small slip, see Section 5.1) at nearly constant rotational speed. Configurations consisting of pumps and PDSs are used for operation at variable speed.

The corresponding modes of operation are called *fixed-speed operation* and *variable-speed operation*, respectively, in the context of EEI. The same reference flow-time profile applies to both fixed-speed operation and variable-speed operation.

In the case of pumps operated at fixed speed in variable flow systems, the variation of the flow rate is normally realized by variable throttling, for example by the means of a control valve.

The four possible combinations of constant or variable flow systems and fixed or variable speed operation are shown in Table 8.1.

It is obvious that variable-speed operation in constant flow systems is not reasonable from an economic point of view because the higher investment costs of a PDS compared to a motor are normally not compensated by a saving of energy costs. In variable flow systems, fixed-speed operation (with a line-fed AC induction motor) is generally disadvantageous compared to variable-speed operation in respect to electric energy consumption.

In general, a flow-time profile according to Eq. (1.2) is a special one for each individual application. However, for the assessment of pump-drive configurations, common and standardized flow-time profiles shall be used as *reference flow-time profiles* that are described by a limited number of points, not by a continuous curve or function. These discrete points are pairs of values of $Q/Q_{100\%}$ and $\Delta t/t_{tot}$.

TABLE 8.1 Operation Modes of Pump-Drive Configurations

Constant flow system, fixed-speed operation	Constant flow system, variable-speed operation
Variable flow system, fixed-speed operation	Variable flow system, variable-speed operation

In respect to the hydraulic installation, different categories have to be distinguished. These are

- closed loop systems
- open loop systems.

Typical closed loop systems are applications where the purpose of the pump is to transport energy inside a system from an energy supply source to an energy sink. Typical examples are heating systems or cooling distribution systems. Typical open loop systems are applications where the purpose of the pump is to move liquid from one location to another one.

A special type of open loop systems is represented by distribution of water in tall buildings to the various tapping points. For this application, pressure boosting stations serve for supplying the water and for generating sufficient pressure at any tapping point. A special flow-time profile representative for booster stations is proposed and presented in Section 8.3.

The reference flow-time profiles that shall be applied for the determination of EEI of pump-drive configurations used in constant or variable flow systems are given in Tables 8.2 and 8.3.

The value $Q_{100\%}$ has to be defined individually for each category of pump-drive configurations (for circulators see Section 8.2; for single pump units and booster stations see Section 8.3).

The reference flow-time profile for variable flow systems reflects the typical ranges and time fractions of demanded flow rate. Experimental field studies on HVAC (= heating, ventilation, air conditioning) systems, as for example reported in Ref. [9], served as a basis for the definition of the reference flow-time profile for variable flow systems (except booster stations) presented in Table 8.3.

TABLE 8.2 Reference Flow-Time Profile for Constant Flow Systems

Relative flow rate $Q/Q_{100\%}$ (%)	75	100	110
Fraction of operating time $\Delta t/t_{tot}$ (%)	25	50	25

TABLE 8.3 Reference Flow-Time Profile for Variable Flow Systems (Except Booster Stations)

Relative flow rate $Q/Q_{100\%}$ (%)	25	50	75	100
Fraction of operating time $\Delta t/t_{tot}$ (%)	44	35	15	6

8.1.4   Reference Control Curves

In many pumping systems, a control loop serves for automatically controlling a process variable, such as pressure, flow rate, or temperature. The pressure control serves to meet requirements on demanded head and is chosen as the basis in respect to EEI. Control curves that are implemented in automatic control systems are normally described by continuous mathematical functions. For the determination of EEI, representative *reference control curves* shall be used.

For *variable flow systems*, the following (linear) reference control curve is based on experience and on common agreement within Europump:

$$\frac{H}{H_{100\%}} = 0.5 + 0.5 \cdot \frac{Q}{Q_{100\%}} \tag{8.3}$$

The relative values of flow rate and head in Eq. (8.3) are decimal numbers <1.0. Multiplication by the factor 100 yields the relative values in (%). The values $Q_{100\%}$ and $H_{100\%}$ have to be defined specifically for each category of pump-drive configurations (for circulators, see Section 8.2; for single pump units see Section 8.3). For booster stations, a different reference control curve shall be applied (see Section 8.3).

For *constant flow systems*, only the use of pumps operated at fixed speed is reasonable as explained above. The head of a fixed-speed pump is fully determined by the $Q-H$ curve of the pump for constant speed (or more exactly, for constant frequency of the electric supply). Therefore, the reference control curve applied for the determination of EEI, which is valid for constant flow systems, is defined formally to coincide with the $Q-H$ curve of the pump. As a consequence, the EEI determined and declared for variable flow systems will be worse if the respective pump size is only capable to be operated at fixed speed than if it is capable to be operated at variable speed.

8.2   THE CONCEPT APPLIED TO CIRCULATORS

For circulators, the EN Standard [16] and the EU Regulation [22] give information on particular definitions in the context of EEI and on the method to determine actual EEI values of a size.

8.2.1   Particular Definitions

The values of $Q_{100\%}$ and $H_{100\%}$ are defined as flow rate and head, respectively, at the operating point of maximum hydraulic power $P_{\mathrm{hyd,max}}$.[1] For

---

1. The 100% values are defined by the maximum of the hydraulic power because it would be extremely complicated to determine experimentally the best efficiency point (BEP) of the hydraulic part of a circulator. Also—contrary to separate pumps—the nominal data of circulators are defined for the operating point of maximum hydraulic power.

circulators that can be operated at different settings, the values of $Q_{100\%}$ and $H_{100\%}$ belong to the maximum setting curve, which gives the maximum (=rated) hydraulic power. This curve can be either a control curve or a non-control curve, if that exists.

The reference flow-time profile according to Table 8.3 has to be used for the calculation of the weighted average $P_{1,avg}$ of the electric power input.

The reference control curve according to Eq. (8.3) has to be used for standardized measurements and calculation of the compensated power input P_L. (The determination of the compensated power input is described below.)

The reference power $P_{1,ref}$ is defined as a mathematical function and is only dependent on the nominal (=rated) hydraulic power $P_{hyd,r}$:

$$P_{1,ref} = 1.7 \cdot P_{hyd,r} \cdot (1 - e^{-0,3 \cdot P_{hyd,r}}) \tag{8.4}$$

The calculation of EEI is slightly different for different categories of circulators.

For stand-alone circulators and product integrated circulators, EEI[2] is calculated by the equation

$$\text{EEI} = \frac{P_{L,avg}}{P_{1,ref}} \cdot 0.49 \tag{8.5}$$

The factor 0.49 in Eq. (8.6) is a so-called calibration factor. It ensures that at the time of defining the factor only 20% of circulators of the respective type and application have an EEI ≤ 0.20.

For product integrated circulators, developed for primary circuits of solar thermal systems and for heat pumps, EEI is calculated by the equation

$$\text{EEI} = \frac{P_{L,avg}}{P_{1,ref}} \cdot 0.49 \cdot \left[1 - e^{\left(-3.8 \cdot \left(\frac{n_s}{30}\right)^{1.36}\right)} \right] \tag{8.6}$$

Equation (8.7) takes into account an influence of the specific speed n_s. The specific speed is calculated according to Eq. (4.20), but with the values $n_{110\%}$, $Q_{100\%}$, and $H_{100\%}$ instead of the values n_N, Q_{BEP}, and H_{BEP}.

8.2.2 Determination of EEI

The EEI of a size of circulators is determined by tests and evaluations. Important features are summarized as follows.

Test equipment and measuring instrumentation for the flow rate, head, and input power have to meet requirements defined in the ISO Standard [17] for tests according to grade 1.

---

2. According to the standard [16] the original denomination is ε_{EEI}, but it is permissible to substitute the parameter ε_{EEI} by the abbreviation EEI in data sheets, manuals, leaflets, brochures, etc.

The values $Q_{100\%}$ and $H_{100\%}$ are determined by the following steps:

- measuring $Q-H$ values at ≥ 10 operating points grouped around the expected value of $Q_{100\%}$
- approximation of the $Q-H$ curve by a best-fit polynomial function of third degree
- calculation of the function $P_{hyd} = f(Q)$ (according to Eq. (1.15)) by using the best-fit $Q-H$ function
- determination of the flow rate, which corresponds to the maximum of the approximate $Q-P_{hyd}$ curve
- calculation of the value $H_{100\%}$, which corresponds to the value $Q_{100\%}$ using the best-fit $Q-H$ function.

For the value of $H_{100\%}$ determined by this proceeding, the tolerance t is stated to be either -20% or -0.5 m, whatever absolute value is the greater one.

Based on the values $Q_{100\%}$ and $H_{100\%}$, the set points are calculated according to the reference flow-time profile (Table 8.3) and to the reference control curve (Eq. (8.3)).

Besides some further provisions for the tests, clear rules are defined for the adjustment of the flow rate at the set points:

- The directly measured flow rate can be used, if its deviation ΔQ from the value specified by the reference flow-time profile is in the interval $(-5\%$ of $Q_{100\%}) \leq \Delta Q \leq 0$.
- Interpolated values are permitted to be used, if measured values of the flow rate exist within the interval $(-10\%$ of $Q_{100\%}) \leq \Delta Q \leq (+10\%$ of $Q_{100\%})$ around the value defined by the reference flow-time profile; otherwise the next higher value of Q has to be used.

Deviations of the measured head H_{meas} from the head H_{ref} specified by the reference control curve are taken into account by correcting the measured electric power input $P_{1,meas}$ into the so-called compensated power input P_L. The compensated power P_L is calculated in the following way:

- $P_L = (H_{ref}/H_{meas}) \cdot P_{1,meas}$, if $H_{meas} \leq H_{ref}$
- $P_L = P_{1,meas}$, if $H_{meas} > H_{ref}$.

8.3   THE CONCEPT APPLIED TO PUMP UNITS

Configuration consisting of one or several pump(s) and drive(s) that have separate casings are denoted as pump units in this context. The drives can be line-fed motors or PDSs. Pump units that consist of only one pump and its drive are denoted as single pump units. Multiple pump units consist of more than one pump and drive. A special case of multiple pump units are booster stations serving for pressure boosting in buildings.

Pump units that are produced and placed on the market and/or are put into service in a large number are called type series of pump units.

The application of EEI on pump units has not yet reached the status of official EU documents or published standards. All necessary details for a future application have been elaborated in an experimental and theoretical project at the TUD reported in Refs [31,32] and by extensive discussions in the responsible Joint Working Group of Europump. A proposal for standardization, based on this work and its results, is currently under consideration in the responsible Working Group of CEN.

In the first phase, the preparatory work focused on the application of EEI on *single pump units*. Also, current work on standardization is primarily concerned with EEI for this type of pump units consisting of one clean water pump combined with a line-fed motor (for fixed-speed operation) or with a PDS (for variable-speed operation). This comprises pump types and pump attributes in correspondence to the scope of the EU Regulation [24] and of the standard [18].

Also, considerations and preparing work were started by the Joint Working Group of Europump that concern the application of EEI to *booster stations for buildings*. Some particular aspects of this application are presented as an outlook in Sections 8.3.2 and 8.4.9.

Regarding particular definitions for pump units, this chapter presents the current state of consideration at the time this book was written (early 2015).

The application of the concept of EEI to pump units has to take account of the fact that the determination and declaration of the EEI value of pump units is the responsibility of

- a pump manufacturing company that places on the market and/or puts into service the complete pump units
- a company that only assembles the components supplied by different manufacturers to complete pump units and places them on the market and/or puts them into service.

Therefore, methods to determine the EEI of pump units should be practicable in both cases.

Appropriate *experimental* methods and proving their applicability on single pump units in the frame of the project reported in Ref. [31] are shortly described in this chapter.

Similar to the case of MEI (see Chapter 7), the mean EEI_{mean} is relevant for assessing the energy efficiency of a type series of pump units. EEI_{mean} can be determined by tests and evaluations

- on only one individual pump unit out of a type series
- on a sample consisting of a limited number (preferably ≥ 5) of pump units out of the same type series.

Recommendations for the proceeding will be given in the standard, which is currently being considered.

Because of the unavoidable *measurement uncertainties* of all quantities (Q, H, P_1) that need to be measured and of the effect of error propagation in the evaluation procedure, the values of EEI of a tested *individual* pump unit can only be determined to be confined to a respective confidence interval of 95% probability (see the explanation in relation to MEI in Chapter 7).

A semi-analytical model (SAM) for pump units, its validation, and its application for determining EEI is described in Section 8.4. The experimental method as well as the mathematical method are suited for standardization. They could also be incorporated in future legislation.

8.3.1 Application of the Concept of EEI to Single Pump Units

8.3.1.1 Particular Definitions

A *single pump unit* is defined to be a configuration consisting of a glanded pump and a line-fed dry motor or a PDS with a dry motor. An electric motor is a dry motor if the motor rotor rotates in air. In the case of a PDS, the motor and the CDM may be mechanically and/or functionally integrated or non-integrated.

Booster stations that contain only one pump are excluded from the particular definitions and methods of determining EEI for single pump units. Single pump boosters belong to the category booster stations (see the definition of booster in Section 8.3.2.1), and the corresponding particular definitions and methods shall apply for them.

The *values of* $Q_{100\%}$ *and* $H_{100\%}$ are defined as flow rate and head, respectively, at the operating point of maximum unit efficiency $\eta_{\text{unit,max}}$. The unit efficiency is defined by Eq. (1.16). The operating point of maximum unit efficiency is preferred as 100% point instead of that one of maximum pump efficiency (BEP) for two reasons:

- In sample tests on complete pump units, the experimental determination of the efficiency of the incorporated pump would require the determination of the mechanical power P_{mech}; see Eq. (1.14). For this purpose, either the installation of a torquemeter into the unit between drive and pump would be necessary (what would change the mechanical configuration of the unit), or the availability of a calibrated efficiency characteristic of the drive would be needed to enable the calculation of P_{mech} from the measured electric power P_1. Both possibilities are difficult or impossible to be realized, especially by a company that only assembles the unit or by an authority in the frame of verification tests.
- Defining $Q_{100\%}$ and $H_{100\%}$ as the values $Q_{\text{BEP,mean}}$ and $H_{\text{BEP,mean}}$ of the pump size taken from the documentation (e.g., catalogs) of the pump manufacturer would have an additional—and not justified—influence of manufacturing tolerances on the results of *tests on individual pump units*

TABLE 8.4 Reference Flow-Time Profiles and Reference Control Curves for Single Pump Units

Type of System and Mode of Operation	Reference Load-Time Profile	Reference Pressure Control Curve
Constant flow, fixed speed	Table 8.2	Q–H curve of pump
Variable flow, fixed speed	Table 8.3	Q–H curve of pump
Variable flow, variable speed	Table 8.3	Eq. (8.3)
Constant flow, variable speed (not reasonable)	Table 8.2	Eq. (8.3)

if the set points of Q and H would be determined by the mean BEP and not by the BEP of the individual pump.

The *reference flow-time profiles* and the *reference control curves* for determining the EEI for variable and/or constant flow systems are compiled in Table 8.4.

The *reference power* $P_{1,\text{ref}}$ for single pump units shall be defined as the electric power input of a reference pump unit equipped with a reference pump which

- has the nominal data of rotational speed n_N, flow rate Q_{BEP} at $n = n_N$, and head H_{BEP} at $n = n_N$ of the actual pump being part of the unit
- has a reference value of the MEI (according to its definition in Refs [18,24])
- is operated with clean cold water at its nominal operating conditions (n_N, Q_{BEP}, H_{BEP})
- is driven by a reference electric drive which
  - has a nominal mechanical output power, which exactly equals the mechanical power of the pump at its nominal operating conditions
  - performs at exactly the drive efficiency required as a minimum value for a certain efficiency class.

According to the proposal of Europump for a standard on EEI of single pump units, the reference value MEI = 0.4 should be part of the definition of $P_{1,\text{ref}}$.

In previous activities of the Joint Working Group of Europump and in the frame of the respective project at the TUD, see Refs [31,32,42], the reference electric drive being part of the definition of $P_{1,\text{ref}}$ was defined as a three-phase AC induction motor which

- is fed by an electric grid with a frequency of 50 Hz
- is of the two-pole type for the pump nominal speed $n_N = 2900$ 1/min, or is of the four-pole type for the pump nominal speed $n_N = 1450$ 1/min

- has a motor efficiency according to the minimum values for motors of class IE 3 calculated by the polynomial equations provided in the IEC Standard [11].

Other possible definitions, for example to define the reference electric drive as a reference power drive (RPDS) with related losses according to the EN Standard [15], are still being considered.

8.3.1.2 Experimental Determination of EEI

All provisions for the test concerning the pump shall be in accordance with the test Standard ISO 9906 [17], grade 2. The exception for input power of 10 kW and below (as allowed for the application of Ref. [17] on acceptance tests) shall not be valid.

All provisions for the test concerning an electric motor if it is part of the pump unit and is fed directly from an electric grid shall be in accordance with the test Standard IEC 60034-2-1 [12].

All provisions for the test concerning a PDS if is part of the pump unit shall be in accordance with the Standard EN 50598-2 [15].

The values $Q_{100\%}$ and $H_{100\%}$ shall be determined by the following steps:

- measuring the flow rate Q, the pump head H, and the electric power input P_1 at constant motor stator frequency $f_M = f_{grid}$ for a sufficient number of operating points around the expected value of $Q_{100\%}$
- calculation of the unit efficiency η_{unit} by Eq. (1.16) for each test point
- approximation of the $Q-H$ curve and of the $Q-\eta_{unit}$ curve by best-fit polynomial functions of third degree
- calculation of the flow rate $Q_{100\%}$ that corresponds to the maximum of the best-fit $Q-\eta_{unit}$ curve
- calculation of the value $H_{100\%}$ that corresponds to the value $Q_{100\%}$ using the best-fit $Q-H$ function.

Based on the values $Q_{100\%}$ and $H_{100\%}$ the set points are calculated according to the respective reference flow-time profile and—in the case of variable speed operation—to the reference control curve.

Besides some further provisions for the tests, maximum deviations of the adjusted and measured flow rate at the set points shall be within a limit, which has still to be finally defined.

In the case of fixed-speed operation, the measured electric input power $P_{1,meas}$ at the set points is directly taken for the calculation of $P_{1,avg}$.

In the case of variable speed operation, deviations of the measured head H_{meas} from the head H_{ref} specified by the reference control curve shall be taken into account by correcting the measured electric power input $P_{1,meas}$ in the following way:

- $P_{1,corr} = (H_{ref}/H_{meas})^x \cdot P_{1,meas}$, if $H_{meas} \leq H_{ref}$
- $P_{1,corr} = P_{1,meas}$, if $H_{meas} > H_{ref}$.

For the exponent x the following approximate equation was developed and proposed based on simulations with the aid of the SAM described in Section 8.4:

$$x = 1.05 - e^{0.03 \cdot y}; \quad y = \frac{P_{BEP}}{1 \text{ kW}} \tag{8.7}$$

Finally, EEI is calculated from the test results by Eqs (8.1) and (8.2).

8.3.1.3 Investigations in the Frame of the Experimental Project [31]

One task of the project at the TUD reported in Ref. [31] was to prove the method of determining the EEI of single pump units and to investigate some issues of the tests that may have influence on the results.

Therefore, a representative selection of pump units with different individual attributes was assorted and tested in accordance to the method of determining EEI described above.

The selection included pump units that contained pumps of the types ESOB, ESCCi, and MS. The specific speed of the pumps covered the range $12.9 \leq n_s \leq 94.9$ 1/min. The nominal mechanical power of the pumps covered the range $0.9 \leq P_{BEP} \leq 58.9$ kW. The pump units with $P_{BEP} \leq 22$ kW were tested on a test rig in the laboratory of the Chair of Fluid Systems at the TUD, while the pump units with $P_{BEP} > 22$ kW were tested at test facilities of pump manufacturers that are members of the Joint Working Group of Europump. The complete test program comprised variations

- of the electric components (line-fed motor instead of PDS, motors and CDMs of different nominal power and/or supplied by different manufacturers, control mode of the CDM) that were combined with some of the selected pumps
- of properties of the electric cable connecting the CDM and the motor of some of the selected PDSs
- of the inner surface roughness for selected pumps.

Tests were performed according to the reference flow-time profile for variable flow systems (see Table 8.3) and for the reference control curves according to Eq. (8.3) as well as to Eq. (8.8).

As a special feature of the test arrangement and instrumentation, for each tested configuration a torquemeter was installed between the shafts of motor and pump. Additionally, a rotational-speed sensor was installed. This installation made it necessary to modify the mechanical configuration of the tested pump units but enabled it to determine the shaft torque T, the mechanical power P_{mech}, and (separately) the efficiencies η_{pump} and η_{drive}.

For all tests, the random, instrument, and total measurement uncertainties were carefully determined. From these, the total (absolute) measurement uncertainty of EEI resulting from the tests and evaluations was calculated to be in the range from 0.006 to 0.009 and, therefore, smaller than the second decimal of EEI.

Some important results of the investigations are summarized as follows:

- The deviations of the adjusted values of flow rate Q and head H at the set points from the values specified by the reference flow-time profile and reference control curve, respectively, were of a magnitude comparable to the total measurement uncertainties of Q and H.
- From repeated tests on some of the pump units, the reproducibility of the resulting EEI was better than the total measurement uncertainty of EEI.[3]
- The variation of the conductor cross section of the connecting cable[4] in an exemplary test showed no significant effect on the resulting EEI.
- The variation of the connecting cable from 0.5 to 10 m in an exemplary test showed no significant effect on the resulting EEI. Though, a cable of 100 m length worsened the resulting EEI by about 0.014, which value is greater than the magnitudes of the reproducibility as well as the total measurement uncertainty of EEI. As a consequence, a future standard defining the experimental determination of EEI should state an upper limit of the length of the connecting cable used in the tests.
- For two pumps of the type ESOB with a specific speed of about 20 1/min, the inner surfaces showed corrosion after a few weeks of water-filled stand-still on the test rig. The pump efficiency determined before and after corrosion and then again after mechanical smoothing of the corroded surfaces showed differences of about 1.5% (absolute value), which was greater than the typical total measurement uncertainty of the pump efficiency (about 0.7% absolute value).[5] This variation of surface roughness by corrosion is estimated to affect the EEI by about 0.008 to 0.023, which values are greater than the typical total measurement uncertainty as well as the typical reproducibility of EEI. As a consequence, a future standard describing the experimental determination of EEI should state conditions for the test pump in respect to avoid effects of corrosion of inner surfaces. This may be of special importance for verification tests.

---

3. This result does not reflect the effect of the manufacturing tolerance of EEI since the repeated tests were carried out with the same individual pump units.

4. No variation of the conductor cross section below a value given by established engineering practice ("undersizing") was investigated. This is prohibited due to technical reasons besides energy efficiency.

5. The measured differences of pump efficiency caused by varied surface roughness confirm the fundamental explanations in Section 4.1.

8.3.2 Outlook to the Application of the Concept of EEI on Booster Stations

8.3.2.1 Particular Definitions

The following definitions correspond to the current state of considerations in the Joint Working Group of Europump.

Booster means an assembly of one or multiple clean water pump unit(s) together with backflow prevention and additional components influencing hydraulic performance and components necessary to control pressure in open loops inside buildings and which is placed on the market and/or put into service as one single product.

The *values of $Q_{100\%}$ and $H_{100\%}$* are defined by the values indicated on the nameplate and correspond to the maximum hydraulic power $P_{hyd,max}$ of the complete booster station.

The *reference flow-time profile* for booster stations is defined by the values in Table 8.5.

The reference flow-time profile for booster stations reflects the typical range and time fractions of demanded flow rate. It is based on the experience of suppliers and on the study reported in Ref. [10].

The number of discrete points of the reference flow-time profile for booster stations is larger than for variable flow systems in general because those configurations of booster stations shall also be covered, which enable it to adjust the flow rate only in finite steps by on/off switching of pumps.

The *reference control curve* for booster stations is defined by Eq. (8.8):

$$H/H_{100\%} = 0.75 + 0.25 \cdot (Q/Q_{100\%}) \tag{8.8}$$

The relative values of flow rate and head in Eq. (8.8) are decimal numbers <1.0. Multiplication by the factor 100 yields the relative values in (%). The reference control curve is based on investigations and considerations reported in Ref. [10].

TABLE 8.5 Reference Flow-Time Profile for Pressure Boosting Stations in Buildings

Relative flow rate $Q/Q_{100\%}$ (%)	10	20	30	40	50	60	70	80	90	100
Fraction of operating time $\Delta t/t_{tot}$ (%)	6	21	26	19	12	6	4	3	2	1

The *reference power* $P_{1,\text{ref}}$ of booster stations is defined as the electric power input of a reference pump unit with one vertical multistage pump, which

- has the nominal flow rate $Q_{\text{BEP}} = Q_{100\%}$ (defined above), the nominal head $H_{\text{BEP}} = H_{100\%}$ (defined above), a nominal rotational speed $n_N = 2900$ 1/min, and a reference specific speed $n_{s,\text{ref}}$ and a reference value of the MEI (according to its definition in Refs [18,24])
- is operated with clean cold water at its nominal operating conditions (n_N, $Q_{100\%}$, $H_{100\%}$)
- is driven by a reference electric drive that
  - has a "nominal" mechanical power output, which exactly equals the mechanical power of the (virtual) pump at its nominal operating conditions
  - performs at exactly the drive efficiency required as a minimum value for a certain efficiency class.

According to the current state of discussions and agreements within the Joint Working Group of Europump

- the reference value of MEI should be 0.4
- the reference specific speed should be 45 1/min
- the reference electric drive should be a three-phase AC induction motor that
  - is fed by an electric grid with a frequency of 50 Hz
  - is of the two-pole type
  - has a motor efficiency according to the minimum values for motors of class IE 3 calculated by the polynomial equation provided in the IEC Standard [11].

Final definitions of the reference pump and the reference drive to be proposed for standardization are still being considered.

8.3.2.2   Experimental Determination of EEI

The method to determine EEI of a booster station is currently being considered in the Expert Working Group for booster stations of Europump. Some aspects have already been agreed. For example, the experimental determination of EEI shall be performed without (the effect of) a membrane tank on the pressure side of the booster station and in the operation mode of automatic control. Some other questions are still open (e.g., how adjustment deviations and control deviations occurring in the test can/shall be taken into account).

To practically investigate and prove an appropriate test method that can be standardized, a new project at the TUD is currently being prepared.

8.4 SAM FOR PUMP UNITS

8.4.1 Motivation and Basic Principles

For the determination of EEI of individual pump units by tests and evaluations, appropriate test rigs and instrumentation are needed that meet all requirements stated in the respective standards [12,13,15,17]. To determine the mean value EEI_{mean} of a type series of pump units being confined to a confidence interval as small as possible, not only one individual pump unit but a sample of a sufficiently large number (at least five) of pump units out of the same type series has to be tested. This applies also for cases that some modifications in respect to the electric components (e.g., motors and/or CDMs with the same nominal data, but supplied by different manufacturers) exist within a type series, and therefore tests have to be carried out for each modification. These procedures require considerable effort and costs. Especially in the case that components supplied by different manufacturers are assembled to complete pump units by "assembling companies," it is not only time- and cost-intensive but even difficult to hold ready the necessary test equipment and manpower.

On the other hand, data in respect to the energy efficiency of the components (pump, motor, CDM, or PDS) are available, since their determination and documentation are required by standards (e.g., Ref. [15]) and/or the existing EU Regulations [20,24] and can be provided together with the hardware to assembling companies. Particularly, these data are mean values of the respective size.

These data can be used as input for mathematical models that describe the relevant performance and loss characteristics of the components and—by appropriate combination of the component models —for a mathematical model of the complete units. Insofar as these models are not based purely on theoretical correlations and equations, but need input of data that result from tests (or experience), they are called SAMs (= Semi-Analytical Models). The aim of the SAM of a complete pump unit is to enable the determination of the (mean) EEI by only performing calculations without the need to carry out tests on one or several complete pump unit(s).

The method of using a SAM of single pump units is described in detail in the report [32]. Also in the publication [42] some essential issues of the method applied to single pump units were presented. In a more general form the method is described in Part 1 of the standard [15] for applications on extended products (EPs) consisting of motor systems and a driven machine. The method will be an important issue of a standard on EEI of pump units, which is currently being elaborated by the responsible standardization group of CEN.

The mathematical model of a complete single pump unit is composed of part models of the pump and of the electric components. The complete model takes into account the physical interactions of the components (see explanations in Chapter 5).

The model of a pump as part of a pump unit consists of mathematical correlations (=equations) that describe the performance in a general form, which reflects the underlying physical processes and influences but needs a limited number of data (at so-called *supporting points*) which result from tests. These data serve to "calibrate" the general equations. In this sense, the models used for the pump and for the complete pump unit are semi-analytical. On the other hand, the model that describes the losses of a motor or of a PDS (or of an electric motor and of a CDM, which are combined to a PDS) consists of purely mathematical equations that serve to inter-/extrapolate losses in a certain range of load points on the basis of known values of these losses at a limited number of *supporting load points* (specified by the EN Standard [15]). Load points of a motor or of a PDS are assigned by the corresponding values of rotational speed n and torque T.

The models describe the relation between physical input and output quantities (=performance variables) of the components as well as of the complete unit by mathematical equations. These are, the dependence of the pump head H on the flow rate Q and on the rotational speed n or the dependence of the power losses $P_{V,PDS}$ of a PDS on the shaft torque T and on the rotational speed n.

While the result of tests to determine the EEI of an individual pump unit shows generally a (total) measurement uncertainty, the accuracy of any value (especially of EEI) calculated by the means of models is dependent on the magnitude of the *model uncertainties*.

Model uncertainties are the maximum possible deviations (with a probability of typically 95%) of the calculated values of the output variables of the model from the correct values of the output variables of the real component for the same values of the input variables. Their magnitude depends on the assumptions, approximations, and simplifications, which determine the quality and correctness of the models. The actual magnitude of model uncertainty of the models that are described below and that are usable for the calculation of EEI was quantified by validation of the models; see Ref. [32].

For the practical application of the method, an appropriate software tool that only needs as input a very limited amount of data for the isolated components and generates as output the EEI of the pump unit can be developed and then provided to those who are responsible for the determination and declaration of the EEI of pump units.

8.4.2   SAM for Single Pump Units

The model and the method described in this chapter were validated and are applicable for pump units equipped with

- a single pump of one of the types ESOB, ESCC, ESCCi, or MS with nominal data in accordance to the (identical) scope of Refs [18,24]
- a line-fed AC induction motor

- a PDS consisting of an AC induction motor and a CDM with a nominal power $P_{M,N}$ of up to 150 kW.

Other pump types, especially submersible multistage pumps (MSS), are not covered.

8.4.2.1 Model of Pumps

The SAM of pumps serves to convert the hydraulic operating points, given as pairs of values of flow rate Q and pump head H (=input variables), to the corresponding mechanical load points, expressed as pairs of rotational speed n and shaft torque T (=output variables). The latter are input variables of the model of the motor (in the case of fixed-speed operation) or of the PDS (in the case of variable-speed operation) and determine the corresponding shaft power P_{mech}.

The SAM of pumps within the range of applicability is based on

- the approximation of the $Q-H$-characteristic and of the $Q-P$-characteristic, respectively, at the nominal speed $n_{\text{pump,N}}$ by best-fit polynomials of third degree
- the similarity or affinity laws that describe the variation of the operating quantities (flow rate Q, pump head H, shaft power P, shaft torque T, and pump efficiency η_{pump}) if the rotational speed n is varied
- an empirical correlation that describes (small) Reynolds number effects (see Section 4.2) on the operating quantities, especially for variations of the rotational speed n within the relevant range.

Combining the best-fit polynomials at the nominal pump speed $n_{\text{pump,N}}$ with the affinity laws leads to the dimensionless characteristics at any rotational speed n described by the following equations:

$$
\frac{H}{H_{\text{BEP,N}}} = a_{\text{H}} \cdot \frac{\left(\frac{Q}{Q_{\text{BEP,N}}}\right)^3}{\left(\frac{n}{n_{\text{pump,N}}}\right)} + b_{\text{H}} \cdot \left(\frac{Q}{Q_{\text{BEP,N}}}\right)^2 + c_{\text{H}} \cdot \left(\frac{Q}{Q_{\text{BEP,N}}}\right) \cdot \left(\frac{n}{n_{\text{pump,N}}}\right)
$$
$$
+ d_{\text{H}} \cdot \left(\frac{n}{n_{\text{pump,N}}}\right)^2
$$

(8.9)

$$
\frac{P}{P_{\text{BEP,N}}} = k_{\text{corr}} \cdot \left[a_{\text{P}} \cdot \left(\frac{Q}{Q_{\text{BEP,N}}}\right) + b_{\text{P}} \cdot \left(\frac{Q}{Q_{\text{BEP,N}}}\right)^2 \cdot \left(\frac{n}{n_{\text{pump,N}}}\right) \right.
$$
$$
\left. + c_{\text{P}} \cdot \left(\frac{Q}{Q_{\text{BEP,N}}}\right) \cdot \left(\frac{n}{n_{\text{pump,N}}}\right)^2 + d_{\text{P}} \cdot \left(\frac{n}{n_{\text{pump,N}}}\right)^3 \right]
$$

(8.10)

$$\frac{T}{T_{\mathrm{BEP,N}}} = \frac{\left(\frac{P}{P_{\mathrm{BEP,N}}}\right)}{\left(\frac{n}{n_{\mathrm{pump,N}}}\right)} \tag{8.11}$$

In the Eqs (8.9)–(8.11) the relative values of flow rate Q, head H, mechanical power P, shaft torque T, and rotational speed n are decimal numbers. The subscript BEP,N marks the value of the respective operating quantity at the BEP of the pump at its nominal rotational speed $n_{\mathrm{pump,N}}$. Within the relevant range, the following approximate correction factor k_{corr} can be applied:

$$D^2_{\mathrm{impeller}} \cdot \frac{n}{60} \geq 1 \Rightarrow k_{\mathrm{corr}} = 1$$

$$D^2_{\mathrm{impeller}} \cdot \frac{n}{60} < 1 \Rightarrow k_{\mathrm{corr}} = \left(\frac{n}{n_{\mathrm{pump,N}}}\right)^{-\alpha}, \tag{8.12}$$

A suitable value of the exponent in Eq. (8.12) was found in Ref. [32] to be $\alpha = 0.15$. By the determination of the coefficients of the polynomials, the model is "calibrated" to the actual pump. Depending of the number of supporting points that are used for this purpose, two types of the pump model are distinguished:

- For the *type 1* of the pump model, a relatively large number of supporting points (typically 10 or more) that result from measurements of the $Q-H$- and $Q-P$-characteristics at $n = n_{\mathrm{pump,N}}$ serves as input.
- For the *type 2* of the pump model, only three $Q-H$ as well as $Q-P$ supporting points (at 75%, 100%, and 110% of $Q_{\mathrm{BEP,N}}$, respectively) are needed as input. These values are usually available from the qualification procedure in respect to MEI and serve to generate two additional $Q-H$ as well as $Q-P$ supporting points (at 10% and 25% of $Q_{\mathrm{BEP,N}}$, respectively) by the means of empirical correlations.

For the derivation of these empirical correlations, $Q-H$ and $Q-P$ characteristics of a multitude of pump sizes of the types ESOC, ESCCi, and MS published by pump manufacturers in their catalogs and data sheets as well as $Q-H$- and $Q-P$-characteristics of individual pumps of these types determined by the measurements reported in Ref. [31] served as data base. Details and the resulting correlations are described in Ref. [32].

In Figures 8.1 and 8.2, results from application of the pump model, type 2, are compared to measured operating points for two exemplary pumps tested in the frame of the project reported in Ref. [31]. In this case, the $Q-H$ and $Q-P$ values at the three supporting points that serve as input data for the model were taken directly from the measured characteristics of the tested individual pumps. Obviously, the $Q-H$ and $Q-P$ characteristics that result from the application of the model, type 2, show a sufficient agreement with the measurement results.

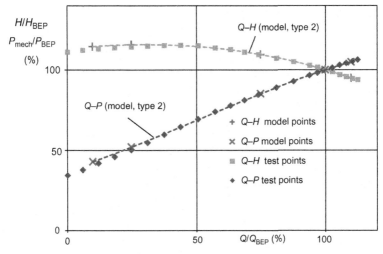

FIGURE 8.1 Comparison of results from test and model application, test pump of the type ESOB, specific speed $n_s = 22$ 1/min.

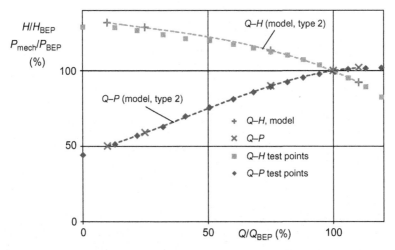

FIGURE 8.2 Comparison of results from test and model application, test pump of the type MS, specific speed $n_s = 39$ 1/min.

8.4.2.2 Model of Line-Fed AC Induction Motors

The model of the motor serves to determine—together with the pump model and via the mechanical equilibrium of motor and pump—the actual values of rotational speed n (caused by slip) and shaft torque T and finally the corresponding electric power input P_1.

As basis for the mathematical model of AC induction motors serve the $T-n$ curve at constant grid frequency f_{grid} (shown schematically in Figure 5.5) and the dependence of the motor efficiency η_M on the relative mechanical power $P_{mech}/P_{M,N}$ (shown for an exemplary motor in Figure 5.3).

As only the nearly linear part of typical $T-n$ curves (between $T = 0$ and $T = T_{M,N}$) is important for operating points of mechanical equilibrium with the driven machine (=a pump), this part is approximated by the linear equation:

$$\frac{T}{T_{M,N}} = \frac{n_{sync} - n}{n_{sync} - n_{M,N}} \tag{8.13}$$

In Eq. (8.13) n_{sync} is the synchronous rotational speed of the motor according to Eq. (5.5), and $n_{M,N}$ is the nominal rotational speed of the motor (i.e., the rotational speed at the nominal motor output power $P_{M,N}$). The relative torque T is a decimal number.

According to the EU regulation [20] for motors, it is mandatory for motor manufacturers to determine and provide in the motor documentation the values of motor efficiency η_M at 100%, 75%, and 50% of the rated load (=$P_{M,N}$). Based on these values, the coefficients of a polynomial of second degree can be directly calculated:

$$\eta_M = a_\eta \cdot \left(\frac{P_{mech}}{P_{M,N}}\right)^2 + b_\eta \cdot \frac{P_{mech}}{P_{M,N}} + c_\eta \tag{8.14}$$

This equation describes the motor efficiency sufficiently well within the range of relative load caused by a pump. The relative mechanical power P_{mech} is a decimal number.

8.4.2.3 Model of PDSs

The model of the PDS serves to convert the mechanical load points, given as pairs of values of rotational speed n and shaft torque T (=input variables that are generated by the pump model and transferred to the PDS model), to the corresponding losses $P_{L,PDS}$ of the PDS, and finally to the corresponding electric input power P_1.

The mathematical model of PDSs is an inter-/extrapolation scheme for the related losses $p_{L,PDS}$ at actual mechanical load points (caused by a pump) in the relevant domain of in the whole characteristic $p_{L,PDS} = f(T, n)$. The basis for the inter-/extrapolation are supporting mechanical load points that are specified in Part 2 of the EN Standard [15]. According to this standard, the related losses at eight specified supporting load points have to be determined and provided in the documentation of a PDS.

Here only the case is treated that values of related losses at the specified supporting load points are available for the complete PDS. The calculation of the related losses of a PDS on the basis of available related values of losses of the motor and of the CDM that are combined to a PDS is described in Part 2 of the Standard [15].

The nominal torque of the PDS is the same as that of the electric motor that is part of the PDS. The nominal torque $T_{M,N}$ of the motor can be calculated from the nominal shaft power $P_{M,N}$ of the motor and the nominal rotational speed $n_{M,N}$. According to Part 2 of the standard [15], the 100% values of rotational speed and torque of the PDS are identical to the respective nominal (= rated) values of the motor. The (mechanical) load points are described as pairs of the relative rotational speed $n/n_{M,N}$ and the relative torque $T/T_{M,N}$. According to Part 2 of the standard [15], the related losses at respective supporting load points defined by the ratio

$$p_{L,PDS\ (i,j)} = \frac{P_{L,PDS(i,j)}}{P_{M,N}} \tag{8.15}$$

are denominated by $p_{L,PDS(i,j)}$. The subscripts in brackets give the relative values $i = T/T_{M,N}$ and $j = n/n_{M,N}$ in (%).

The mechanical load points of pump units that correspond to the reference flow-time profile for variable flow systems (according to Table 8.3) combined with the reference control curve (according to Eq. (8.3)) are located only in a limited domain of the whole field of possible load points of a PDS. Within this domain the related power losses of a PDS can be approximately calculated on the basis of only the three supporting load points $p_{L,PDS\ (100,100)}$, $p_{L,PDS\ (100,50)}$ and $p_{L,PDS\ (50,25)}$ (see Figure 8.3) that are located on the borders of the domain that is relevant for pump units.

Because of the relatively small and nearly linear variation of the related losses within the domain of load points that is relevant for pump units, a simple bi-linear inter-/extrapolation, respectively, is sufficiently accurate for the calculation of related losses within this domain. These inter-/extrapolation equations are the only mathematical equations of the model of the PDS that are needed to determine the EEI of the complete pump unit. It is "calibrated" for the actual PDS by the related power losses at the three supporting load points shown in Figure 8.3.

The inter-/extrapolation equation is:

$$p_{L,PDS} = k_{corr} \cdot \left[B_n \cdot \left(\frac{n}{n_{M,N}} \right) + B_T \cdot \left(\frac{T}{T_{M,N}} \right) + C \right] \tag{8.16}$$

The three coefficients in Eq. (8.16) can be calculated directly from the known values of the related power losses at the three supporting load points. The borderline between the interpolation and extrapolation domains is described by the Eq. (8.17):

$$\frac{T}{T_{M,N}} = 0.5 \cdot \frac{n}{n_{M,N}} \tag{8.17}$$

Within the interpolation domain, the correction factor k_{corr} in Eq. (8.16) is equal to 1.0. For the extrapolation domain, based on the model validation

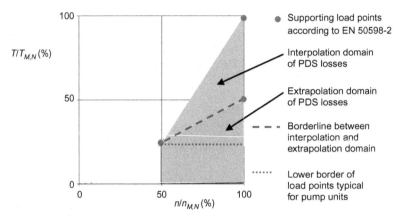

FIGURE 8.3   Relevant domains of load points in the $T-n$ diagram of PDSs.

reported in Ref. [32] an empirical correlation was derived for the correction factor k_{corr} in the form

$$k_{\text{corr}} = \left(\frac{(T/T_{M,N})}{0.5 \cdot (n/n_{M,N})} \right)^{-\beta} \qquad (8.18)$$

A suitable value of the exponent in Eq. (8.18) was found in Ref. [32] to be $\beta = 0.25$. In the equations (8.16) to (8.18), the relative values of torque T and rotational speed n are decimal numbers.

It has to be noted that the inter-/extrapolation scheme described here differs slightly from that one described in Ref. [15]. However, it was proved to be sufficiently accurate for pump units by results of measurements on several PDSs in the frame of the experimental project reported in Ref. [31].

8.4.2.4   Determination of EEI by Application of the SAM

The physical interactions between the components of pump units are described in Sections 5.1 and 5.2 (see especially Figures 5.1 and 5.6). These interactions are represented mathematically in the SAM of complete pump units.

For single pump units that consist of a pump and of an AC induction motor, the physical equilibrium of pump and motor in respect to the rotational speed n and the shaft torque T for any value of the flow rate Q is mathematically represented by setting equal the (absolute) pump torque T_{pump} (described by Eqs (8.10) and (8.11)) and the (absolute) motor torque T_M (described by Eq. (8.13)). Regarding the absolute values of torque, the different nominal values of pump and motor torque that serve as reference for the respective relative torque values have to be taken into account.

Because of only small deviations of the actual rotational speed n from the nominal rotational speed of the motor (caused by the slip s) the correction factor in Eq. (8.10) can be set approximately to 1.0 in this case.

The resulting equation is a cubic equation for the rotational speed in dependence on the relative flow rate $Q/Q_{BEP,N}$, which can be solved arithmetically by appropriate mathematical methods. With the resulting value of the rotational speed, for each value of the relative flow rate $Q/Q_{BEP,N}$

- the mechanical power P_{mech} can be calculated by Eq. (8.10)
- the motor efficiency η_M can be calculated by Eq. (8.14)
- the electric input power P_1 results as the ratio of P_{mech} and η_M as the output of the complete model.

For single pump units that consist of a pump and a PDS, the pump model serves for converting the hydraulic load points (given as pairs of the relative flow rate $Q/Q_{BEP,N}$ and of the relative pump head $H/H_{BEP,N}$) to the corresponding mechanical load points (described as pairs of the relative torque $T/T_{BEP,N}$ and of the relative rotational speed $n/n_{pump,N}$). This is executed by transforming Eq. (8.9) into a cubic equation for the relative rotational speed $n/n_{pump,N}$ that corresponds to the respective load point. This cubic equation can be solved arithmetically by appropriate mathematical methods.

The values of the relative rotational speed $n/n_{pump,N}$ resulting from the solution of the cubic equation are inserted into Eqs (8.10) and (8.11) that deliver the corresponding values of the relative mechanical power $P_{mech}/P_{BEP,N}$ and of the relative torque $T/T_{BEP,N}$ at the respective hydraulic load point.

For the determination of the losses in the PDS with the aid of the PDS model, the mechanical load points have to be converted to the relative rotational speed and relative torque related to the nominal values of the electric motor $n_{M,N}$ and $T_{M,N}$, respectively:

$$\frac{n}{n_{M,N}} = \frac{n_{pump,N}}{n_{M,N}} \cdot \frac{n}{n_{pump,N}} \tag{8.19}$$

$$\frac{T}{T_{M,N}} = \frac{T_{BEP,N}}{T_{M,N}} \cdot \frac{T}{T_{BEP,N}} \tag{8.20}$$

In equations (8.19) and (8.20) the relative values of torque T and rotational speed n are decimal numbers. For the mechanical load points that result as output of the pump model, the related power losses $p_{L,PDS}$ of the PDS are calculated by Eqs (8.16) and (8.17). The corresponding values of the absolute power loss of the PDS and of the electric input power result from equations

$$P_{L,PDS} = p_{L,PDS} \cdot P_{M,N} \tag{8.21}$$

$$P_1 = P_{mech} + P_{L,PDS} \tag{8.22}$$

The determination of EEI for single pump units consisting of a pump and an AC induction motor by application of the SAM of the complete unit is schematically illustrated in Figure 8.4.

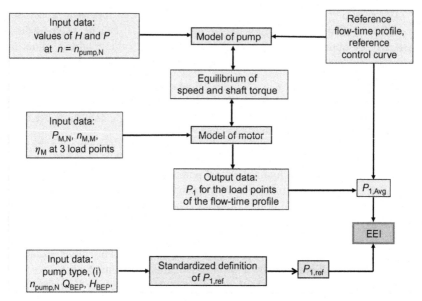

FIGURE 8.4 Determination of EEI by application of the SAM of pump units for fixed-speed operation (schematic representation).

The determination of EEI for single pump units consisting of a pump and a PDS by application of the SAM of the complete unit is schematically illustrated in Figure 8.5.

As input for the models of the pump and of the drive, only some nominal data and a very limited number of values (in the case of the pump model, type 2, only six values, for the motor model or for the PDS model only three values) at supporting points are needed. The SAM of the complete unit generates as output for all hydraulic load points—defined by the respective reference flow-time profile and (for variable flow systems) by the reference control curve—the electric input power P_1 and its weighted average value $P_{1,\text{avg}}$ according to the definition Eq. (8.2).

For both categories of pump units, the first step is the determination of the values $Q_{100\%}$ and $H_{100\%}$. For this purpose, for five values of the flow rate Q (grouped around the expected value of $Q_{100\%}$) the corresponding values of the head H and P_1 are calculated by the means of the respective SAM. With the values of Q, H, and P_1 the corresponding values of the unit efficiency η_{unit} are calculated by Eq. (1.16) and a best-fit approximation function $\eta_{\text{unit}} = f(Q)$ is determined. The mathematically determined maximum of this function yields the corresponding values $Q_{100\%}$ and $H_{100\%}$.

With the values of $Q_{100\%}$ and $H_{100\%}$ the specific hydraulic load points of the respective pump unit are determined, and the value of $P_{1,\text{avg}}$ is generated by the SAM.

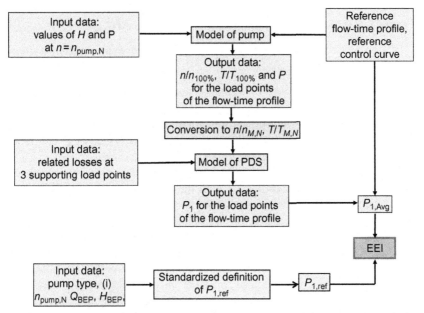

FIGURE 8.5 Determination of EEI by application of the SAM of pump units for variable-speed operation (schematic representation).

The reference power $P_{1,\text{ref}}$ is calculated according to the definition given in Section 8.3. The EEI of the respective pump unit results as the ratio of $P_{1,\text{avg}}$ and $P_{1,\text{ref}}$.

8.4.2.5 Validation of the SAM of Complete Pump Units

The determination of EEI by application of the SAM was validated by results of the tests reported in Ref. [31] for the case "variable flow systems" in combination with "variable speed operation." The details of the validation are described in Ref. [32].

For the pump model, type 1, all operating points of the $Q-H$ and $Q-P$ characteristics measured at $n_{\text{pump,N}}$ in the tests were taken as basis for the best-fit approximation Eqs (8.9) and (8.10).

For the pump model, type 2,

- first the $Q-H$ and $Q-P$ supporting points at the "MEI points" of the tested pumps were determined from the tests according to Chapter 7
- then the additional supporting points at 25% and 10% of Q_{BEP} were calculated by using the empirical correlations mentioned above
- and finally the approximation equations for the $Q-H$ and $Q-P$ characteristics were determined as best-fit functions on the basis of the five supporting points each (see exemplary Figures 8.1 and 8.2).

TABLE 8.6 Estimated Maximum (Relative) Difference ΔEEI_{max}

	For Pump Model, Type 1	For Pump Model, Type 2
ΔEEI_{max}	$\pm 2.9\%$	$\pm 3.2\%$

To avoid effects of adjustment deviations in the tests, the output value $P_{1,avg}$ of the SAM was calculated using the *measured* values of Q and H at the hydraulic load points from the tests.

The same value of the reference power $P_{1,ref}$—determined according to its general definition in Section 8.3 with the individual values of Q_{BEP} and H_{BEP} resulting from the tests—was used to calculate EEI_{test} and EEI_{model}. The relative differences ΔEEI (as percentage of EEI_{test}) of the calculated EEI_{model} (= output of the SAM of the complete pump unit) and the corresponding experimentally determined value EEI_{test} can be understood as a measure for the quality of the SAM. Though, it is important to note that these differences are not identical to model uncertainties because they also include the measurement uncertainties of the experimentally determined values. In fact, the true values of EEI are only known from the test to be confined to the corresponding confidence interval (of 95% probability), whose width is determined by the total measurement uncertainty of EEI_{test}. As reported in Section 8.3, the absolute measurement uncertainty of EEI_{test} was in the range of $0.006-0.009$ for the test results that served for comparison with EEI_{model}.

For a total of nine tested pump units out of the test program reported in Ref. [31], the values of ΔEEI and their standard deviation $\sigma_{\Delta EEI}$ from the arithmetic mean $(\Delta EEI)_{mean}$ were determined. According to the statistical mathematics, the maximum values of ΔEEI can be expected—with a probability of 95%—to be within the range $\pm 1.96\, \sigma_{\Delta EEI}$. These values are presented for both types of the pump model as part of the SAM of the complete pump unit in Table 8.6.

The actual model uncertainty of EEI_{model} is expected to be slightly smaller if pump model type 1 is applied than if pump model type 2 is applied. However, this small advantage of application of pump model, type 1, must be contrasted with the considerably greater effort of necessary tests (type 2 is based on already existing results from the test procedure in respect to MEI).

8.4.2.6   Use of the SAM for Sensitivity Studies

The SAM of complete pump units is not only a useful tool for the determination of EEI. It can also be used to study and to quantify the effects of particular tolerances and uncertainties on the total tolerance of values of EEI that are determined by the methods described in Sections 8.3 and 8.4. Such sensitivity studies were performed in the frame of the project reported in Ref. [32]. Two examples of sensitivity studies and corresponding results are shortly presented here.

As explained in Chapter 6, the manufacturing performance tolerances of the components of pump units have an influence on the manufacturing tolerance $(t_{EEI})_{man}$ of values of EEI that result from tests or from the application of the SAM. Assuming that the manufacturing performance tolerances of the components of the pump unit are statistically independent of each other, $(t_{EEI})_{man}$ can be estimated by applying the statistical law of error propagation ($=$ square root of the sum of squares).

In the case of pump units consisting of a pump and a motor operated at constant frequency (*fixed-speed operation*), the manufacturing tolerance of EEI results as:

$$(t_{EEI})_{man} = \sqrt{(t_P)^2_{man} + (t_{\eta_M})^2_{man}} \tag{8.23}$$

In the case of pump units consisting of a pump and a PDS (*variable speed operation*) the manufacturing tolerance of EEI results as:

$$(t_{EEI})_{man} = \sqrt{(F_{H \to P_{1,avg}})^2 \cdot (t_H)^2_{man} + (F_{P \to P_{1,avg}})^2 \cdot (t_P)^2_{man} + (F_{pL \to P_{1,avg}})^2 \cdot (t_{pL})^2_{man}} \tag{8.24}$$

In these equations,

- $(t_x)_{man}$ designates the manufacturing tolerance of the respective performance quantity x. The subscript P refers to the mechanical power of the pump, and the subscript pL refers to the related losses of the PDS.
- $F_{x \to y}$ are sensitivity factors that quantify the effect of the manufacturing tolerance of the performance quantity x on the manufacturing tolerance of the performance quantity y.

The manufacturing tolerance of $P_{1,avg}$ is equivalent to the performance tolerance of EEI as the reference power $P_{1,ref}$ is defined by nominal data that are not affected by tolerances.

For the case *variable flow systems, variable speed*, simulations were carried out using the SAM of the complete unit.

To simulate performance tolerances of the pump and of the PDS, examples were calculated by assuming

- constant tolerances $(t_H)_{man}$ and $(t_P)_{man}$, respectively, applied to the whole $Q-H$ and $Q-P$ characteristics
- a constant tolerance $(t_{pL})_{man}$ applied to the related losses $p_{L,PDS}$ at each of the three supporting load points.

From these simulations, the sensitivity factors in Eq. (8.24) were determined. By inserting their numerical values, Eq. (8.24) becomes:

$$(t_{EEI})_{man} \approx \sqrt{1.2 \cdot (t_H)^2_{man} + (t_P)^2_{man} + 0.04 \cdot (t_{pL})^2_{man}} \qquad (8.25)$$

In Table 8.7 values of $(t_{EEI})_{man}$ are presented for combinations of exemplary values of performance tolerances of the components (pump and PDS) of pump units.

Also by application of the SAM, the sensitivity of the tolerance of EEI_{test} determined by tests on measurement uncertainties was studied.

The uncertainty of the electric input power P_1 at any load point of a test results from

- measurement uncertainties of the flow rate Q, of the measured electric power input $P_{1,meas}$, and—in the case of variable speed operation—of the pump head H
- the uncertainty of the values of $Q_{100\%}$ and—in the case of variable speed operation—of $H_{100\%}$ (resulting from the determination of the maximum of the $Q-\eta_{unit}$ curve)
- adjustment deviations of the flow rate Q
- the uncertainty of the correction of the measured electric power input $P_{1,meas}$ by Eq. (8.7).

For each of these uncertainties, corresponding sensitivity factors were quantitatively determined by application of the SAM of the complete unit. By assuming

- values of the uncertainties that are plausible from experience or are actually specified by standards
- that the various uncertainties are statistically independent of each other and the statistical law of error propagation can be applied

the values of $(e_{EEI})_{meas}$ (=measurement uncertainty of EEI) presented in Table 8.8 result.

The magnitude of the measurement uncertainty $(e_{EEI})_{meas}$ indicated in Table 8.8 for the case *variable speed, variable flow* is comparable to the actual values determined for the tests reported in Ref. [31].

8.4.3   SAM for Booster Stations—An Outlook

In principle, the method to determine EEI by application of the SAM of complete pump units can be extended to multiple pump units, for example to booster stations. Since booster stations consist of one or several single pump units(s) in combination with an appropriate control unit, the already existing SAM of single pump units can serve as a basis. The SAM of complete booster stations must additionally incorporate

TABLE 8.7 Manufacturing Tolerance $(t_{EEI})_{man}$ for Exemplary Values of Performance Tolerances of the Components of Pump Units

	$(t_P)_{man}=0\%$ $(t_{pL})_{man}=0\%$	$(t_P)_{man}=0\%$ $(t_{pL})_{man}=\pm10\%$	$(t_P)_{man}=\pm3\%$ $(t_{pL})_{man}=0\%$	$(t_P)_{man}=\pm3\%$ $(t_{pL})_{man}=\pm10\%$
$(t_H)_{man}=\pm3\%$	$(t_{EEI})_{man}=\pm3.3\%$	$(t_{EEI})_{man}=\pm3.9\%$	$(t_{EEI})_{man}=\pm4.5\%$	$(t_{EEI})_{man}=\pm4.9\%$
$(t_H)_{man}=0\%$	$(t_{EEI})_{man}=0\%$	$(t_{EEI})_{man}=\pm2.0\%$	$(t_{EEI})_{man}=\pm3.0\%$	$(t_{EEI})_{man}=\pm3.6\%$

TABLE 8.8 Estimated Measurement Uncertainty $(e_{EEI})_{meas}$

Fixed Speed		Variable Speed, Variable Flow
Constant Flow	**Variable Flow**	
±2.0%	±1.8%	±2.7%

- the resulting characteristics of two or more single pump units that are arranged and operated in parallel
- additional hydraulic losses caused by piping and valves that are part of the complete unit
- the effects of the control unit (on/off switching points of individual pumps in the case of multiple pumps, adjustment deviations, and control deviations).

The development and validation of an extended SAM for booster stations are also envisaged in the frame of the new project at the TUD (see outlook in Section 8.3), which is currently being prepared.

8.5 CONCLUDING REMARKS

The importance of reducing the consumption of electric energy by pumping systems is undoubted and is the focus of corresponding standardization and legislation. A prerequisite for achieving the intended amount of energy saving by setting efficiency related requirements consists in appropriate methods to assess the energy efficiency of electric motor driven pump units. Concerning pump units that consist of separate pumps and drives, existing EU Regulations and International Standards are directed—up to now—on pumps and electric motors as individual products. A weakness of assessment methods described in these documents is their focusing on the product efficiency at nominal or near nominal operating conditions. Though, very many applications of pumps are characterized by wide ranges of demanded flow rate and head. This fact justifies the introduction, definition, and use of the EEI as a valuable indicator for assessing the energy efficiency of complete configurations that consist of pumps and their drives. For circulators that are integrated pump-motor configurations, the assessment of their energy efficiency by EEI is already well established, is required by an EU Regulation, and is described in a corresponding EN Standard.

Also for pump units that consist of separate pumps and drives and that can be considered as EPs (Extended Products), EEI is a preferable indicator for the assessment of their energy efficiency. Necessary definitions of EEI for pump units are being elaborated by Europump and can serve for future standardization and legislation. Besides a method for determining the EEI by performing and evaluating tests on individual specimens of complete pump units, a mathematical method based on a SAM of complete single pump units was developed and validated. The application of this method needs as input only some data of the components (pump and motor or PDS) that are available and can also be provided by component manufacturers to assemblers. This method is also capable of being extended to multiple pump units, especially to booster stations for buildings. Therefore, the method is proposed to be included in an EN Standard on EEI of pump units, which is currently being considered. The mathematical method should have a good chance of being accepted as a tool for assessing the energy efficiency of pump units as EPs in future legislation.

References

[1] Eurostat, Tables, Graphs, and Maps.

[2] Strommix in der EU 27, Entwicklung der Stromerzeugung in Europa von 2007 bis 2030 (engl.: Mix of electricity in EU 27, development of electricity generation in Europe from 2007 to 2030), edited by VDMA; 2010.

[3] Wagner H-J, Koch M-K, et al. CO_2-Emissionen der Stromerzeugung (engl.: CO_2 emissions of electricity generation). BWK 2007;59(10):44−52.

[4] Bertoldi P, Atanasiu B. Scientific and Technical Reports Electricity consumption and efficiency trends in European Union—Status Report 2009. European Commission Joint Research Center (JRC); 2009.

[5] Waide P, Brunner CU. Working Paper Energy-efficiency policy opportunities for electric motor-driven systems. International Energy Agency; 2011.

[6] AEA Energy & Environment (Author: Falkner H.): Appendix 6, Lot 11, Pumps: (in commercial buildings, drinking water pumping, food industry, agriculture), Report to the European Commission; 2008.

[7] AEA Energy & Environment (Author: Falkner H.): Appendix 7, Lot 11, Circulators in buildings, Report to the European Commission; 2008.

[8] De Almeida AT, et al. EuP Lot 11 Motors, Final report to the European Commission; 2008.

[9] Hirschberg R. Annual European part load profile of heating pumps, Technical Report VDMA Pumps + Systems; 2001.

[10] Hirschberg R. Efficiency evaluation of booster stations using load profiles and control curve, Technical Report VDMA Pumps + Systems; 2014.

[11] IEC 60034-30: 2008: Rotating electrical machines—Part 30: Efficiency classes of single-speed, three-phase, cage-induction motors (IE-code).

[12] IEC 60034-2-1: 2007: Rotating electrical machines—Part 2−1: Standard methods for determining losses and efficiency from tests (excluding machines for traction vehicles).

[13] IEC/TS 60034-2-3: 2013: Rotating electrical machines—Part 2−3: Specific test methods for determining losses and efficiency of converter-fed AC motors.

[14] IEC 60034-31: 2010 (Ed.1.0): Rotating electrical machines—Part 31: Selection of energy-efficient motors including variable speed applications—Application guide.

[15] EN 50598: 2014: Ecodesign for power drive systems, motor starters, power electronics & their driven applications.

[16] EN 16297: 2013: Pumps—Rotodynamic pumps—Glandless circulators.

[17] ISO 9906: 2012: Rotodynamic pumps—Hydraulic performance acceptance tests—Grades 1, 2 and 3.

[18] prEN 16480: 2013: Pumps—Minimum required efficiency of rotodynamic water pumps.

[19] Directive 2005/32/EC of the European Parliament and of the Council of July 6, 2005 establishing a framework for the setting of eco-design requirements for energy-using

products and amending Council Directive 92/42/EEC and Directives 96/57/EC and 2000/55/EC of the European Parliament and of the Council, recasted through Directive 2009/125/EC of the European Parliament and of the Council of October 21, 2009 establishing a framework for the setting of ecodesign requirements for energy-related products.

[20] European Commission Regulation (EC) No. 640/2009 implementing Directive 2005/32/EC of the European Parliament and of the Council with Regard to Ecodesign Requirements for Electric Motors, (2009), amended by Commission Regulation (EU) No 4/2014 of January 6, 2014.

[21] Working Document on possible requirements for electric motors and variable speed drives, Draft Ecodesign Regulation.

[22] Commission Regulation (EC) No 641/2009 of July 22, 2009 implementing Directive 2005/32/EC of the European Parliament and of the Council with regard to ecodesign requirements for glandless standalone circulators and glandless circulators integrated in products, amended by Commission Regulation (EU) No 622/2012 of July 11, 2012.

[23] Working Document on a possible Commission Regulation amending Commission Regulation (EC) No 641/2009 of July 22, 2009 with regard to ecodesign requirements for glandless standalone circulators and glandless circulators integrated in products.

[24] Commission regulation (EU) No 547/2012 of June 25, 2012 implementing Directive 2009/125/EC of the European Parliament and of the Council with regard to ecodesign requirements for water pumps.

[25] Commission communication 2012/C 402/07 in the framework of the implementation of Commission Regulation (EU) No 547/2012 implementing Directive 2009/125/EC of the European Parliament and of the Council with regard to ecodesign requirements for water pumps; 2012.

[26] ENER Lot 28—Pumps for private and public wastewater and for fluids with high solids content, Final report to the European Commission; 2014.

[27] ENER Lot 29—Pumps for private and public swimming pools, ponds, fountains, and aquariums (and clean water pumps larger than those regulated under ENER Lot 11), Final report to the European Commission; 2014.

[28] EuP Lot 30—Electric motors and drives, Final report to the European Commission; 2014.

[29] Industry Commitment—To improve the energy performance of stand-alone circulators through the setting-up of a classification scheme in relation to energy labeling, Europump; 2005.

[30] Roth M, Ludwig G, Bischof V. Final report to Europump A method to define a minimum level for pump efficiencies based on statistical evaluations. TU Darmstadt; 2007.

[31] Lang S, Ludwig G. Final report to Europump Experimental investigations on variable-speed single pump units with asynchronous motors. TU Darmstadt; 2014.

[32] Stoffel B. Final report to Europump Development, validation and application of a Semi-Analytical Model for the determination of the Energy Efficiency Index of single pump units. TU Darmstadt; 2014.

[33] Guideline on the application of COMMISSION REGULATION (EU) No 547/2012 implementing Directive 2009/125/EC of the European Parliament and of the Council with regard to ecodesign requirements for water pumps, Europump Guide; 2012.

[34] Extended Product Approach for pumps, Europump Guide (Draft version of October 27, 2014).

[35] Stoffel B, Lauer J. Summary of the final report on the research project for Europump Attainable efficiencies of volute casing pumps. TU Darmstadt; 1998.

[36] Attainable efficiencies of volute casing pumps, Europump Guide No. 2; 1999.

[37] Stoffel B, Ludwig G, Meschkat S. Evaluation of efficiency values considering the effect of pump size modularity. TU Darmstadt; 2002.

[38] European guide to pump efficiency for single stage centrifugal pumps; 2003.

[39] Stoffel B, Lauer J. Summary of the final report on the research project for VDMA Theoretically attainable efficiency of centrifugal pumps. TU Darmstadt; 1994.

[40] Study on improving the efficiency of pumps, Report produced for the European Commission—SAVE; 2001.

[41] IEC 60034-1: (Ed. 12.0) 2010: Rotating electrical machines—Part 1: Rating and performance.

[42] Lang S, Ludwig G, Pelz PF, Stoffel B. General methodologies of determining the Energy Efficiency Index of pump units in the frame of the Extended Product Approach. Rio de Janeiro: EEMODS; 2013.

Index

Note: Page numbers followed by "*f*" and "*t*" refer to figures and tables, respectively.

Printed in the United States
By Bookmasters